Multicriteria Methodology for Decision Aiding

Nonconvex Optimization and Its Applications

Volume 12

The titles published in this series are listed at the end of this volume.

Multicriteria Methodology for Decision Aiding

by

Bernard Roy
LAMSADE, Université Paris-Dauphine

Translator

Mark R. McCord
The Ohio State University, U.S.A.

KLUWER ACADEMIC PUBLISHERS
DORDRECHT / BOSTON / LONDON

A C.I.P. Catalogue record for this book is available from the Library of Congress.

ISBN 978-1-4419-4761-1

Published by Kluwer Academic Publishers,
P.O. Box 17, 3300 AA Dordrecht, The Netherlands.

Kluwer Academic Publishers incorporates
the publishing programmes of
D. Reidel, Martinus Nijhoff, Dr W. Junk and MTP Press.

Sold and distributed in the U.S.A. and Canada
by Kluwer Academic Publishers,
101 Philip Drive, Norwell, MA 02061, U.S.A.

In all other countries, sold and distributed
by Kluwer Academic Publishers Group,
P.O. Box 322, 3300 AH Dordrecht, The Netherlands.

Printed on acid-free paper

Table of Contents

Reader's Guide

This book has been organized to help individuals with different backgrounds and motivations find passages of interest more easily. In this way, the material should be accessible to directors and managers, as well as to decision makers or project supervisors and to most consultants or researchers that conduct professional studies. These individuals may be engineers, operations researchers, computer scientists, statisticians, economists, business administrators, or anyone interested in developing an understanding of problems and methods related to decision aiding. The book has also been organized to appeal to researchers wishing to develop this "science."

A systematic reading of the book from beginning to end is not discouraged, but the margin indicators, the cross-references to preceding or following passages, the index, the chapter summaries, and the remarks introducing sections permit a less structured reading that can be conducted according to the interests of the reader. Twelve reference examples directly extracted from real applications are modified to motivate, illustrate, and develop concepts, procedures, and methods. In some sense, each example can stand alone. Two tables (in Chapters 6 and 12) summarize their characteristics and how they are used in the course of the book.

Passages bordered by the dark band in the margin are considered essential to understanding the concepts presented and should not be skipped. Passages that appear in smaller type treat connections or remarks of lesser importance, enter into further details of a specific point, present complex justifications or proofs, add supplementary illustrations or explanations, or cover specific procedures or methods.[1] Readers not wishing to cover the material in depth can skip these passages and still follow the rest of the book. Passages in normal type that are not bordered by the dark band in the margin cover notions, results, examples, developments, and methodological aspects that are not necessarily essential when first encountered. However, the reader may later want to refer to these passages (with the assistance of the extensive cross-referencing used) to understand more fundamental ideas better.

Indications	Reading Level
in the margin	Level 1: Indispensable for understanding what follows
Normal type without margin indicator	Level 2: Can be skipped in the course of the first reading, but returning to the passage may be required
Small type without margin indicator	Level 3: Can be skipped without hindering comprehension (illustrations – details)

[1] *translator's note: In the interest of expediency, we have not translated many of these passages in this version. We indicate in the text where they can be found in the original, French version.*

FOREWORD FOR THE ENGLISH-LANGUAGE EDITION

The aim of this book is to explain the bases of a general decision-aiding methodology which took shape toward the end of the 1960s. At that time I had been working for twelve years as an Operations Researcher at SEMA(Metra International), a consulting firm for which I had become scientific director. Having been confronted with a great variety of concrete problems during this period, I was eager to use these experiences to extend the concepts, analytical techniques, procedures and the significance of results with a view to conceptualizing a methodology that would be as all-inclusive as possible. During the 1970s my appointment to the University of Paris-Dauphine as Professor of Operations Research allowed me to make significant progress in fundamental research in this area. A number of related factors also combined at that time to create a rich and stimulating environment in which this project could grow to maturity: regular meetings with researchers and practitioners from many different countries (primarily from the European Working Group on Multicriteria Aid for Decisions); grants from my own university, as well as from the French National Research Council (CNRS), that enabled me to set up and develop a university research center for Analysis and Modelisation of Decision Aid (LAMSADE); and, finally, consulting work which I undertook for several large corporations and government agencies, particularly the RATP,[1] responsible for public transport in the Ile-de-France region.

All of this is to say that the concepts explained in the present work are the fruits of the reflection and effort of a vast community – principally Europeans, but also French-speaking Canadians. This community forms the core of what was initially known as the French-speaking School, but which is today called the European[2] School of Decision Aiding.

The present work was conceived in the hopes of making a modest contribution to a decision-aiding science. The object of such a science is not to discover or to approximate the best possible decision, but to develop a corpus of conditions and means on which we can base our decisions in light of what we believe to be most suitable. I believe that contributing to this science consists of seeking to develop a network of concepts, models, procedures and results so as to form a structured and coherent body of knowledge. This knowledge – in conjunction with the corpus of hypotheses – can serve as keys to guide decision-making and to facilitate relevant communication in conformity with the decision-maker's objectives and values. Rigorous concepts, well formalized models, precise calculation procedures (notably optimization procedures), and

---

[1] *Régie Autonome des Transports Parisiens.*

[2] *For the origins and characteristics, as well as recent works emanating from this school, see Roy and Vanderpooten (1996).*

axiomatic results should be at the heart of such a science. Through them, we should be able to enlighten and scientifically assist decision-making processes especially by:

- making that which is objective stand out more clearly from that which is less objective;
- separating robust from fragile conclusions;
- dissipating certain forms of misunderstanding in communication;
- avoiding the pitfall of illusory reasoning;
- emphasizing, once they are understood, incontrovertible results.

The difficulties I encountered at the begining of my career as an operations researcher, and later as a consultant, made me realize that there were some limitations on objectivity in decision-aiding. In my opinion, five major aspects must be taken into consideration:

1) The borderline (or frontier) between what is and what is not feasible is often fuzzy. Moreover, this borderline is frequently modified in light of what is found from the study itself.

2) In many real-world problems, the "decision maker D" does not really exist as a person truly able to make a decision. Usually, several people (actors or stakeholders) take part in the decision process, and it is important not to confuse the one who ratifies a decision with the so-called decision maker in the decision aiding process. This decision maker is in fact the person or the set of persons for whom or in the name of whom decision aiding effort is provided.

3) Even when the decision maker D is not a mythical person, his or her preferences rarely seem well-formed. In and among areas of firm convictions lie hazy zones of uncertainty, half-held beliefs or, indeed, conflicts and contradictions. We have to admit, therefore, that the study itself contributes to eliminating questions, solving conflicts, transforming contradictions and to destabilizing certain convictions. If, within this perspective, we decide or accept to resort to a multicriteria approach, the elaboration of a family of criteria cannot be founded on purely objective considerations.

4) Data such as the numerical values of evaluations or performance measures, the characteristics and analytical forms of probabilistic distributions, the weights of criteria, etc., are often imprecise, uncertain, or ill-determined. This is true, for instance, when a cost or a ratio is viewed as a Gaussian random variable, the Gaussian distribution being used afterwards for computing an expected value of the cost or the ratio.

5) In general, it is impossible to say that a decision is a good or a bad one by referring solely to a mathematical model. The organizational, pedagogical and cultural aspects

of the entire decision process that lead to a given decision will also contribute to its quality and success.

As I endeavored to demonstrate in a recent paper,[3] these limitations are compatible with the conception of a decision aiding science outlined above. They underline, moreover, the foundations of what a decision science should be. By decision science, I mean a science whose purpose would be the search for objective truths in decision making and, more particularly, the knowledge – if not precise, then at least approximate – of the best decision within a given context, through the use of models presented as simplifications of reality. Such a science could not, in my view, exist without a first and fundamental postulate which I have termed the postulate of the optimum. This postulate can be formulated as follows:

Postulate of the optimum: In situations likely to involve decision making, there will be at least one optimal decision, namely a decision for which it is possible (on the condition that we have adequate time and resources) to establish objectively that a clearly better decision does not exist. It should be possible to do this while remaining neutral in terms of the decision-making process itself.

Objectivity, which goes hand-in-glove with neutrality, is based here on the following two hypotheses:

Hypothesis 1: A criterion giving meaning to the concept of optimum can be defined independently of any opinion, conviction, value or human prejudice.

Hypothesis 2: The optimal decision can be discovered or approximated and recognized as such independently of the models and procedures used to arrive at it.

In order to accept these two hypotheses, we must also accept another postulate which I have formulated as follows:

Postulate of reality of the first order[4]: The principal aspects of reality (an individual's preferences, the borderline between possible and impossible, the consequences of an action) on which decision aiding is based relate to objects of knowledge. These can be seen as given (existing outside any modelling) and as sufficiently stable (with respect to time, diversity of actors, discourse held, observations made) to enable us to refer to the exact state or the precise value (which can be of either a certain or a stochastic nature) of those specific characteristics deemed significant of one aspect of reality.

We can observe that a decision science, seen in the terms presented above, should not be confused with another science whose purpose is focused on describing and studying how actors decide, even if the purpose of this other science is to develop models that

---

[3] *See Roy (1993).*

[4] *With reference to Watzlawick's (1984) terminology.*

account for actors' behavior and predict some of their decisions (see, for example, Bell *et al.*, 1988). We should observe that this science of decision-making behavior covers a sphere whose concerns are rather distinct from those encountered in operations research or in decision aiding. This does not mean, however, that the body of knowledge it produces cannot contribute to the development of a decision-aiding science.

The European School of Decision Aiding endeavors to account for the limitations on objectivity discussed above and, as a result, to free itself from the postulates and hypotheses I have described as the necessary foundation stones for a decision science. This is probably the way in which the European School differs most distinctly from other currents of thought, and in particular from those known as MCDM and MAUT. A brief description of the European School would state that those who contribute to developing it pay special attention to the six aspects or concerns listed below:

1) A way of proceeding that seeks to structure a problem through the concepts of action and coherent criterion family capable of facilitating communication within the decision process.

2) A conception of decision aiding that proceeds by progressively forming a conviction rather than by discovering an optimum of a pre-existing system of preferences.

3) A way of taking into account the imprecise, the uncertain and the ill-determined, primarily through different types of thresholds, but also through fuzzy or probabilistic modelling.

4) A way of conceiving of or concretely capturing the importance that should be given to each criterion; it is important to go beyond the rather naive concept of weight and to make use of newer approaches for capturing and quantifying the notion of importance, according to the type of aggregation techniques considered.

5) A refusal to look upon the presence of intransitive or incomparable features in certain preference models as irrational; this leads us to use these very features as a means of taking into account a desire not to take a position on certain value systems or as a means of revealing ambiguities, which could be, for example, the consequence of imperfect knowledge.

6) A distinction between the output of a calculation procedure and the recommendation which should be articulated on the basis of a robustness analysis that uses different results from several sets of data.

Any decision will include components of discovery, reasoning, irrational randomness and, finally, organizational effects. It will also develop over the course of a process that progressively reduces the actors' margin of freedom to act. As I have pointed out in the introductory part of this book, scientific decision aiding cannot be considered in the absence of the context in which the decision process unfolds. The actors' positions and behavior are the result of a multiplicity of rationalities, based on different, and sometimes conflicting, systems of values. These positions and behavior arise from the specific points of view within which the actors assume themselves to be judged and from their divergent, even contradictory, perceptions of reality – the result of seeing a situation from different standpoints. All of this allows us to see the real interest of a methodology based on elaborating multiple criteria.

It is nonetheless important to deepen our understanding of the concept, to specify how and on what bases criteria are to be developed, to study the role criteria can play in clarifying the comparison of actions when we do not wish to use the postulate of reality of the first order. This is the aim of the central part of the book. In previous works I have attempted to define what these criteria apply to and in what situations it is conceivable to use them, even if the postulate of the optimum ought to be rethought. It is thus only in the final part of the present work that I broach the problem of how to clarify decisions by advantageously using a family of criteria. At no point in all of the above is there any need to refer to a third postulate, on which many research projects (particularly under the multicriterion decision making label MCDM) have been based, albeit not always explicitly. I would formulate[5] this third postulate in the following way:

Postulate of the decision-maker: Every decision is the act of a decision-maker: a well-identified, powerful actor, acting with reference to a system of rational preferences in the sense of a certain body of axioms that exclude both ambiguity and incomparability and that decision aiding does not seek to modify.

Some readers may regret not finding in the present work precise descriptions of the choice, sorting and ranking algorithms that we developed and experimented with from the earliest versions of ELECTRE methods. These readers can refer to Bana e Costa (1990) or Vincke (1992). A justification, as well as a detailed presentation of many techniques illustrated through case studies can be found (in French) in Roy and Bouyssou (1993). Maystre *et al.* (1994) provides additional applications, as well as practical advice on how to implement certain methods developed within the methodology presented here.

In this book I have attempted to present a conceptual framework and, through very concrete examples, to illustrate how this framework can be used both to obtain results and to formulate recommendations. My hope is that the book will contribute to the overall intellectual development, both of readers interested in research or consulting and of those readers with only a casual interest in operations research.

Going from French to English and adapting the book for an audience that might be loosely described as "Anglo-Saxon" in cultural background, we (the author and the translator Mark McCord) have encountered three major difficulties, which we hope to have overcome. The first of these was of a linguistic nature. I have always considered that choosing the right word is essential, not only in order to communicate, but also in

---

[5] *Cf. Roy (1985), avant-propos.*

order to stimulate thought and encourage further study. In addition, words chosen by any given individual carry with them a certain conception of the relationships which that person maintains with reality. Challenging questions about the three postulates cited above is, of course, strongly bound up with this kind of relationship. This is why I consider the choice of each term to be critical, yet extremely difficult. This choice should attempt to reconcile the need for clear and correct communication with that of producing precise, rigorous statements that unfailingly unite concepts, procedures, results and recommended practices. The examples related to decisions concerning numerous and varied aspects of business activities were chosen precisely in order to facilitate communication with the reader. Yet these examples inevitably refer to customs, habits and norms that are inextricably linked to a specific culture. The second difficulty arose from certain differences between the French culture, in which the examples provided in the book are rooted, and the presumed culture –in fact, cultures – of the English language reader. Finally, the French language edition, although now ten years-old, is not out-dated. In the area of methodology, very little has changed over the past ten years. Nonetheless, it was indispensable to include (or at least to cite) a number of more recent works wherever they could complete or illustrate the methodology presented here. The talent of the translator made these adaptations possible.

It is my hope that this book will enable a wide audience of both practitioners and researchers to become acquainted with and benefit from the methodology described in these pages, a methodology which constitutes a theoretical or intellectual framework directed towards formulating recommendations for action. This framework is still not widely known – and is often misunderstood – in the English-speaking world.

Bernard Roy
Paris, March 1995

Acknowledgments

This book would never have seen the light of day without the work, comments and encouragement of many people from a number of different countries. Although it is impossible to name each of these contributors individually, I would like to express my deepest gratitude and thanks to all of them here.

It is appropriate, first of all, to acknowledge the pioneering work carried out in the 1960s by the research group operating under my direction at SEMA(Metra International). At this time, operations research approached decision problems through the concept of the single criterion whose optimization was presumed to enable researchers to discover the optimal decision. Three decades ago, claiming to reason choices on scientific bases by taking into account clearly identified multiple criteria seemed a very strange idea indeed. In both Europe and America, these were only a handful of research groups working in this direction. The backing, as well as the research work and ideas, of Patrice Bertier, Raphaël Benayoun, Eric Jacquet-Lagrèze, Oleg Larichev, Hubert Le Boulanger, Jean de Montgolfier, Gilbert Sussmann, and several other colleagues, provided the powerful support necessary for me to advance along this path. These men have all, in different ways, undeniably contributed to the methodology presented in this book.

The structuring, testing and further development of this methodology has greatly benefited from the presentations and discussions, held twice yearly since 1975, by the European Working Group on Multicriteria Aid for Decisions. The analyses, remarks and suggestions put forward by numerous friends from the working group have been invaluable. I would especially like to mention Jean-Pierre Brans, Jean-Claude Vansnick and Philippe Vincke, as well as Anna Ostanello, Benedetto Matarazzo, Alain Schärlig, Heinz-Michaël Winkels and many others. In the early 1980s, when this methodology was in the full flower of its first development, I was fortunate enough to attract several brilliant young doctoral candidates to the LAMSADE[1] Research Center, which I had recently set up at the University of Paris-Dauphine. Among them were Denis Bouyssou, Jean Siskos, Jean Moscarola and Daniel Vanderpooten, each of whom has since gone on to establish an international reputation in his own area of speciality. As advances Ph.D. students they performed many an invaluable service. By reading and commenting on the successive versions of this book's first French-language edition and by articulating thought-provoking criticisms of certain passages, they encouraging me to go more thoroughly into the ideas presented here. In some instances, they were able to provide more satisfactory responses than my own to certain difficult questions. My most profound gratitude and heart-felt thanks go out to all of these colleagues.

I am likewise deeply indebted to Dominique François, who single-handedly typed, corrected and re-corrected all the versions of the original French-language manuscript. Special thanks should also go to Ursula McCord for typing the first draft of the English-

[1] *Laboratoire d'Analyse et Modélisation de Systèmes pour l'Aide à la Décision.*

language version. It was on the basis of this first draft that Dominique François took over the final formatting of the present text, for which she had to compose a great diversity of mathematical formulae, complicated figures and tables. Throughout, she has unfailingly demonstrated a remarkable degree of patience, efficiency and attention to detail, which I admire and for which I most gratefully thank her. I would also like to offer my thanks to Dominique Champ-Brunet for her able assistance in up-dating the bibliography for the English-language edition.

As I pointed out in the Foreword to this book, there are many problems involved in translating a work of this type from French to English. In order to overcome these difficulties, I urged the translator, Mark McCord, to make changes in the original text wherever he found it necessary, either for reasons of divergent cultural traditions or in order to bring certain passages up-to-date. He has accomplished this rather daunting task with great sensitivity, as was witnessed in his numerous incisive questions and remarks, proof of the translator's constant effort to avoid meaningless literal renderings while retaining the spirit of the original text. The great care he has taken with the translation allows me to think that the present edition of this book will faithfully transmit the contents of the original French-language version to a non-French speaking audience. It was a great pleasure for me to work with Mark over a period of several years. My contact with him has led me to a better understanding of why and how one might conceive of decision aid in ways other than we are accustomed to within the European School. For all the work we did together and for all these exchanges of points of view, I am profoundly grateful.

My grateful thanks also go out to Lucien Duckstein and Roman Slowinski, whose advice and criticism were of great help in putting together the English text. The former, with his wide-ranging familiarity with both European and American cultures and the latter, with his experience of translating the same book into Polish, have both, through their thorough knowledge of the subject, significantly contributed to surmounting the numerous difficulties involved in perfecting the final translation into English. If, in spite of all these combined efforts, the reader still finds certain passages to be obscure, we can only assume that the obscurity lies not with the translation, but with the original French text.

INTRODUCTORY CHAPTERS

HOW TO AID WHOM WITH WHAT TYPES OF DECISIONS

Chapter 1

DECISION PROBLEMS AND PROCESSES

SUMMARY

We analyze the concept of decision and show that it cannot be completely separated from that of decision process. We then propose that the set of critical points in the course of a decision process determines the comprehensive or final decision. This comprehensive decision results from the interactions among the stakeholders (individuals, entities, communities) and the conflicts among the preferences of the actors (stakeholders, third parties, ...). Finally, we introduce an example concerning the purchase of a family car to illustrate the chaotic and often unforeseen course that the process may take.

Decisions are made when choosing to do or not to do things, or when choosing to do them in certain ways. They may be made at a national or local level; in a company, factory, or department; or even within the family. They may relate to objectives of growth plans, policies for regional development, implementation strategies for high speed train service, modification of existing welfare systems. Or they might consider sites for a new factory, components of an advertising campaign, funding of research products, levels of quarterly dividends, acceptance of loan applications, buying or selling of stocks, hiring of personnel, ...

These decisions are rarely made by a single individual such as a government official, a company president, or a director of a specific department. Even if responsibility for the decision does ultimately rest with a well-identified individual, the decision will generally be the product of an interaction between this individual's preferences and those of others. In many cases, the final decision might not be the responsibility of or influenced by single **individuals**. It could involve what we shall call **entities**: an elected or appointed body, cabinet officials, a board of directors, a personnel department, a labor union, a panel of experts, an admissions committee. It could also involve a group with less well-defined boundaries: a professional lobby, company employees, public opinion, ... In this case we shall use the term **community**.

These **actors** (individuals, entities, communities) are what we shall call **stakeholders** (see Banville *et al.*, 1993), in that they have an important interest in the decision and will intervene to directly affect it through the value systems which they possess. In addition, there are those (citizens, taxpayers, the elderly, company personnel, university students, consumers, ...) who do not actively participate in shaping the decision, but who are affected by its consequences and whose preferences must be considered when arriving at the decision. Based on Sfez (1973), we shall use the expression **third parties** to denote those falling in this category.

In reality, the comprehensive decision develops in a somewhat chaotic manner, evolving from on-going confrontations among the preferences of the different actors (stakeholders, third parties, ...[1]). These confrontations are brought about by parallel and successive interactions among the stakeholders within the framework of their operating environment. **It is the playing out of these confrontations and interactions, under the various compensating and amplifying effects of the system, that makes up what we shall call the decision process**.

In the types of problems mentioned above, the decision may boil down to a single action taken by the one responsible for executing the final choice. The action would usually result from a period of individual reflection and group interaction. This classic view may not hold in many cases, however. The period preceding the final action is often full of intermediate options which may seem like "fragments" of or constraints on the decision. Under these conditions, what constitutes the final course of action is really only a minor part of what is truly the comprehensive decision. Moreover, this final action is frequently just a ratification of previously made decisions. In other cases, the decision might be decomposed into a hierarchy of partial decisions which together make up the comprehensive decision. This comprehensive decision, then, is a synthesis of an interconnected web of decisions. For all of these reasons, **the concept of a decision cannot be completely separated from that of a decision process**.

It will, therefore, be useful to think of the decision as unfolding within the framework of a process whose progress is punctuated by a certain number of critical points, one of which is the final action. These critical points are not necessarily predetermined; nor are they always organized in a logical manner. Rather, the existence, contents, and relative position (following or occurring at the same time as others) of the critical points are greatly affected by the various stakeholders, some of whom will act upon them in an indirect attempt to influence the decision. This feature, as well as many others related to the progress of a decision process, is clearly brought out in Sfez (1973) in the context of two problems related to public transportation in the Paris region. **In any case, it is the set of these critical points in the course of a decision process that determines the comprehensive decision**. We use the term comprehensive decision to avoid confusion with the fragments of this comprehensive decision that are outcomes of the intermediate critical points or with any other partial decisions that the intermediate critical points might generate.

As an illustration, consider the possible process leading to a family's decision on what type of car to buy to replace its current automobile, which is too old and too small for its needs. The family consists of a father, mother, and two children (a 13-year old daughter and a 15-year old son). The father's (one stakeholder's) objectives are clear: The new car must be able to comfortably transport the family between its apartment and country residence each week and be able to make a long trip once a year; he also decides to set an upper limit B on the purchase price. Under these conditions, he readily eliminates from consideration a certain number of models that are too small, too expensive, do not hold the road well enough, ... After this initial screening he considers

---

[1] *In Section 2.2, we shall discuss the role of two other actors (the analyst and client) who are generally different from the stakeholders and third parties.*

a certain number of models as still being feasible. A discussion with his wife leads him to add a model that he had forgotten and to rule out another that would not be adequate for local errands and that, moreover, she considers ugly. His daughter begs him not to forget model a_f, which her best friend's family drives and which is "fantastic." Everyone looks into the pros and cons of the different models retained for consideration. The father focuses his attention on comfort, safety, and operating costs. Given the importance he attaches to these three criteria, he is convinced that model a_1 is the best, followed closely by model a_2. He considers model a_f as being a distant third. He then tries to convince his family that a_1 is the model they should buy. The rest of the family objects in that they believe that the aesthetic qualities of the car should be considered: None of the other three intervening parties likes the looks of a_1. They all believe that a_2 is at least as good looking as a_1. In addition, the children point out that a_1 is not big enough for them to bring friends to their country house. In fact a_2 has only one disadvantage compared to a_1: It is slightly more expensive. The mother and the children list all the advantages of a_2 and point out that these would easily justify the increased cost. Since the price of a_2 is still less than B, the father agrees.

The family is, therefore, on the verge of buying a_2, in spite of the daughter's efforts to persuade everyone that a_f is the best choice, when the son comes in waving an advertisement: A well-known company is bringing out a new model a_n that, according to the advertisement, encompasses the best features of a_1, a_2, and even a_f, without having their disadvantages. The father looks over the advertisement, discovers that the purchase price could be greater than B, and starts to reweigh the pros and cons... The family begins debating once again... The increased safety and good gas mileage of a_n are sufficient reasons for the family to wait for better estimates of the purchase price, and it decides to keep the old car for a few more months. We leave it to the reader to complete the story, which might lead to the purchase of a_n, a_2, or some other model.

Beginning in Chapter 3, we shall see examples that are more in line with the professional nature of this book. This simple example, however, which we shall refer to as **the example of the family car**, illustrates the primary concerns of those who have something at stake in a decision to be made. In general, stakeholders in the process will normally try to push the decision in a desired direction[2] by:

- listing or, more accurately, developing a large number of possibilities to serve as objects of the decision considered;
- analyzing the consequences of each of the possibilities so as to understand the advantages and disadvantages of each;
- comparing the valuations that result in light of the objectives so as to develop a personal conviction of the relative worth of the different possibilities or, at least, of some of the possibilities;

---

[2] *This assumes a certain form of rationality on the part of the stakeholders, which is not always observed in reality but which they would expect from the analyst acting as their "advocate."*

- trying to convince other stakeholders of the direction's merits, so that the evolution of the process conforms as much as possible to the value system underlying these feelings.

These are the concerns of this book. The reader may find it interesting to compare them with those developed in French (1986), Steuer (1986), von Winterfeldt and Edwards (1986), Bell *et al.* (1988), Seo et Sakawa (1988), Tabucanon (1988), Bana e Costa (1990), Edwards (1992), Goicoechea *et al.* (1992), Keeney (1992), Pomerol and Barba-Romero (1993). Here, we present a systematic approach that each stakeholder should find valuable, since it is based on concepts and methods that recognize the importance of the unexpected and even inconsistent events that inevitably influence the way in which a real decision process unfolds. Each stakeholder can expect this approach to produce: clearer and more pertinent organization of data; more scientific means of analysis; more rigorous inferences; and more persuasive arguments for a resulting position.

Chapter 2

DECISION AIDING: MAJOR ACTORS AND THE ROLE OF MODELS

SUMMARY

Before describing how this book approaches the idea of decision aiding, we must examine the idea of a model. We define a model in Section 2.1 and discuss the types of models treated in this book – namely, conscious models possessing some explicit form.

We define decision aiding in Section 2.2 as a model-based activity designed to answer questions posed by some stakeholders in the decision process. The answers sought should clarify the decision and help identify behavior that will increase the compatibility of the process with the stakeholders' objectives and value systems. We analyze the roles of three major actors – the decision maker, the analyst, and the client – and discuss the issues of neutrality and objectivity in the modeling effort.

Consider a decision process concerning investment or production, marketing or finance, distribution or procurement, machine or personnel management. The stakeholder – who may come from a firm, from a community group, from a government administration, ... – will automatically be faced with a certain family of questions putting him or her in contact with a particular class of phenomena. In some cases, simple observation is all that he or she needs to get a good handle on these phenomena and work effectively with their causes and effects. In other cases, however, more formal models can be extremely useful in understanding and working with these phenomena or in developing appropriate answers and convincing others to accept them. As with all scientific approaches, decision aiding relies heavily on relatively explicit and formal models.

2.1 MODELS AND REALITIES

2.1.1 Definition

Grappling with a certain reality requires some preliminary concepts. The subsequent analysis and the information that becomes available will cause others to emerge. All of these concepts can be used as building blocks for a somewhat explicit and rigorous framework, a framework that is called the model.

DEFINITION 2.1: A ***model*** *is a schema*[1] *which, for a certain family of questions, is considered as a representation of a class of phenomena that an observer*[2] *has more or*

---

[1] *An (interiorized) mental or figurative (diagrams, mathematical formulae, ...) description generally reduced to essential features and having a symbolic character.*

[2] *Certain (often implicit) aspects of the relationships between the observer and the system observed (class of phenomena × family of questions) cannot be separated from the model. The model, therefore, is as much a representation of these aspects as it is of the system, a point that should be recognized by anyone using the model as a representation of the class of phenomena for the specific family of questions.*

less carefully removed from their environment to help in an investigation and to facilitate communication.

In some sense, any attempt at observation, reflection, analysis, or persuasion is based on an underlying model. Consider, for example, a discussion about the desirability of studying a possible highway location. Suppose that one of the discussants argues for not studying the specific location because of technical difficulties that would lead to high costs or low capacity. These arguments would result from a (interiorized) model that she used with some degree of rigor to convince herself that the location's foreseeable benefits would not compensate for these and perhaps other unstated difficulties.

As another example, consider the refinement of a questionnaire designed to produce useful information on loan requests received by a bank. Omitting certain types of information and including others, or phrasing the questions in certain ways, implies a certain reliance on a schematic representation of the (perhaps vague) relationships between the possible responses and the reality that they are supposed to depict. In addition, any inferences made from the responses would only have meaning within the framework of an even more complex model.

In trying to develop an organized vision of a class of phenomena that will help provide answers to a family of questions, one might wish to develop an implicit mental representation of the phenomena. This would allow freer use of intuition in arriving at responses. On the other hand, one may wish to make this representation as rigorous as possible, basing it exclusively on facts, quantitative data, and logic. The models addressed in this book cover a range of attitudes between these two extremes and are sensitive to Trémolières' (1975) three dimensions of knowledge: "objective, instrumental, and logical knowledge, called scientific knowledge; sensory or psychosensory knowledge, producing desire and pleasure or fear and sorrow; evocative knowledge leading to comprehensive meaning and unifying partial experiences, ..." (translated quotation). Although sensitive to these dimensions, our discussion will be limited to conscious models possessing some explicit form. While this form need not be mathematical, or even completely specified, when we use the word "model," it will be understood that its form is sufficiently structured to facilitate communication.

It is useful to mention a few points essential to the modeling activity and to the relation between models and realities.

2.1.2 Limiting the scope of the model

A model can only relate to a fragment of reality. In general, one can consider this fragment as a functioning system that can be isolated in a manner consistent with its intended purpose. The fragment of reality is, therefore, specified both by the way it relates to a certain class of phenomena and by the purpose of the family of questions to be addressed. How can an appropriate fragment be isolated? How can it be tied back into the context from which it was carefully but arbitrarily taken?

Consider, for example, the father who had to rely on a certain amount of experience when isolating from the many automobiles those he wished to consider further. This initial list had to be modified because of input from the other family members. Similarly, the limit B on the purchase price could not be considered hard and fast: The family could compensate for a higher price by reducing other purchases or by having one member look for supplemental income. The father excluded these aspects from his rough model; yet the new make of automobile a_n might force him to reconsider the way in which he modeled reality in considering a limit on the purchase price.

Even with the assistance of proven methodologies, which we shall discuss in the following chapters, the professional modeler will not be able to escape a trial and error process. Indeed, the possibilities for determining the model boundaries are overwhelming. To carve out a piece of reality in a useful way, then, will require a good deal of careful observation, imagination, and experience. This is why limiting the scope of the model must be considered an art. There are no procedures that are purely scientific, objective, and independent of the modeler (i.e., the observer) which can be used to distinguish among phenomena and separate important from unimportant ones. Even so, general, mathematical, or model-based systems theory (Klir, 1972; von Bertalanffy, 1973; Wymore, 1976; Le Moigne, 1977; Chapman *et al.*, 1992) can be useful in guiding observations and limiting the numerous options.

2.1.3 The family of questions

Whether the modeling task is concerned with defining subsystems, differentiating among groups, specifying variables, forming structural relations, deciding whether to omit a certain feature, ..., the freedom offered by reality is often disconcerting. The family of questions to be addressed will usually guide and justify the choices made at this level.

The criteria that the father considered in making his comparisons would undoubtedly have been different if the question were one of buying a car for professional use. He would also have reasoned differently if he were to have considered the following question: What would be the effect on the choice of the family car if the price structure (gasoline, tires, insurance, taxes, ...) changed significantly? In addition to affecting cost directly through price per mile, such a change might affect the way in which the car is used, leading to a change in the relative importance attached to the various criteria.

2.1.4 The model as representation

The model is a schema considered as a representation of a class of phenomena. It is, in some ways, a caricature (where the word is meant to be devoid of any negative connotations) of the considered fragment of reality. This caricature could somewhat faithfully represent some features, while playing down or exaggerating some others. That is, the terms "true" and "false" do not apply, *per se*, to a model. We say, rather, that a model is or is not well-suited or pertinent to a given problem.

This representation, this model, serves as an intermediary whose purpose is essentially one of understanding, mastering,[3] reasoning about, and communicating reality. The person studying, using, or developing a model often wonders about the limits of its meaning.

The distinctions made by Régnier (1966) are appropriate: "Given that, on the one hand, the abstract object is entirely constituted by its definition and that, on the other hand, the concrete object can never be exhaustively described, we shall say: an abstract object is a model of a concrete object when the definition of the former is taken as a representation of the latter. I shall call real models those that are constructed to help describe the real structure of the concrete object and nominal models those that help us represent the object as it appears in the experiment" (translated quotation). Even though this distinction was made in the context of a specific domain, it is useful in all sciences, and specifically in the framework of this book. Indeed, the distinction is of the greatest importance for putting the modeler's intentions, even pretensions, into context. Our aim in this book concerns only nominal models.

2.2 DECISION AIDING

2.2.1 Definition

The human ability to represent phenomena abstractly and the aptitude to reason hypothetically and deductively have long been used in the domain of action: People think before they act; they conceptualize before they implement. It is this activity of deduction and modeling that, when consciously performed in an attempt to clarify the behavior of an intervening party in the decision process, comprises the essence of decision aiding. Decision aiding is then defined here as an activity, and the rest of this chapter is devoted to studying the conditions under which this activity is or can be put into practice.

DEFINITION 2.2: ***Decision aiding*** *is the activity of the person who, through the use of explicit but not necessarily completely formalized models, helps obtain elements of responses to the questions posed by a stakeholder of a decision process. These elements work towards clarifying the decision and usually towards recommending,[4] or simply favoring, a behavior that will increase the consistency between the evolution of the process and this stakeholder's objectives and value system.*

---

[3] *This word should not be construed to imply any notion of dominance or supremacy. Rather, we are referring to a function whose intent is to guide an evolution in a voluntarist and reasoned way that conforms to certain end goals. Depending on the case, mastering could signify forecasting, projecting, automating, computerizing, maintaining, watching over, directing, optimizing.*

[4] *Both the analyst and the decision-maker are aware that the decision-maker is completely free to behave as he or she sees fit after the recommendation is made (see Chapters 4 and 6).*

We wish to emphasize that decision aiding is only remotely related to a "search for the truth." As we shall show throughout this book, the theories, methodologies, and models that the analyst may call upon are almost always of a different nature. They are designed to help think through the possible changes that a decision process may facilitate so as to make it more consistent with the objectives and value system of the one for whom, or in the name of whom, the decision aiding is being practiced. These theories, methodologies, and models are meant to guide actions in complex systems, especially when there are conflicting viewpoints. They concern phenomena that are generally difficult to isolate, i.e., to think of outside of a specific context. It is, therefore, difficult to compare these theories, methodologies, or models based on predictions or on tests designed to falsify[5] them. We shall not be particularly concerned, however, with whether or not there exist objective means to judge the "validity"[6] of a given theory, methodology, or model. More subjective arguments will usually suffice, and we shall avoid referring to approximations of some truth.

Throughout this book, then, one must not lose sight of the fact that **decision aiding is meant to assist in constructing, establishing, and arguing for convictions. The basis and the means of developing the decision must be the object of critical discussion**. The methodology presented in this book (concepts, techniques, results, methods, ...) is primarily destined to play this role (see Roy, 1994). If there exists such a thing as "science of decision aiding," it must be considered from this perspective and not from one concerned with demonstrating the optimality of a decision or with dictating the decision to be taken.

2.2.2 Aiding for whom?

Usually, it is one of the stakeholders who is being aided. The various stakeholders in the process might be relatively diverse, having different objectives and conflicting value systems. Therefore, a specific application of decision aiding will rarely be comprehensive enough to benefit all of them. For this reason, decision aiding will almost always require that a particular stakeholder (individual, entity, or community) be identified. We shall call this party the **decision maker**.

---

[5] *In the sense of this term used in Popper (1979).*

[6] *One can find a discussion of this question in Déry et al. (1993), Landry and Oral (1993), Miser (1993), and in Roy (1993). These works consider the position of the positive current, which judges the validity of a model according to its predictive ability rather than the acceptability of its axiomatic foundation as a working hypothesis. This position, which may prove to be useful in some scientific and economic fields, seems more questionable in decision aiding, as will be argued in the rest of this chapter, as well as in Sections 10.3.1 and 11.1.1.*

Identifying a decision maker will entail specifying the objectives under which she[7] operates. This stakeholder generally plays a critical role in the evolution of the process, and it is on her behalf or in her name that the decision aid is applied. It is also possible, however, that this stakeholder is the spokesperson for third parties, even if her intervention in the evolution of the process would not appear at first sight to be critical. The decision aid may be even less "personalized," being based on a consensus of all or some of the stakeholders and third parties. Even in these cases, we shall continue to speak of a decision maker to denote the rather mythical figure defined by a set of objectives commonly held or assumed to be commonly held. The frequent cases in which the modeling procedure requires more detailed specification of value systems can lead to special difficulties here. "Good" and "bad," and even the distinction between what is "possible" and what is not, will rarely have absolute meaning and usually require some discussion. A vaguely defined decision maker will preclude such discussion.

The decision maker, then, is the one who assesses the "possible" and the objectives, and who expresses preferences and has an interest in imposing them on the evolution of the process. As we shall see later, however, this does not imply that the decision aid will exclude opinions, strategies, or preferences of the other stakeholders.

2.2.3 Aiding by whom?

When the decision maker is an individual, she might develop the decision aid herself. When concerned with her own problems, however, the situation would be similar to a doctor having herself as a patient. Moreover, the decision maker may not have the background to perform the decision aid. The one performing the aid is, therefore, generally different from the decision maker. Whether distinct from the decision maker or not, we shall call this individual **the analyst**.

We wish to distinguish our usage of this term from its usual one, which gives the impression of someone looking at the problem and the decision process from the outside. Our experience has indicated that this is rarely the case (see Sections 2.2.5, 2.2.6). Therefore, we emphasize that our analyst (who may also be called the designer) not only looks at the problem and the decision process, but he[8] also influences them.

The analyst is usually an expert or a specialist (systems design engineer, systems analyst, operations researcher, economist, statistician, financial advisor, ...) who works alone or leads a team. He may have had frequent dealings with the decision maker in the past. Or the analyst may be a total stranger to the decision maker, coming from a consulting firm or a functional division of a company.

---

[7] *translator's note: To make the writing less cumbersone and to facilitate comprehension, we shall refer to the decision maker, actors other than the analyst, and general actors in the feminine form and to the analyst in the masculine form. This choice was made randomly.*

[8] *translator's note: see footnote 5.*

Among other things, the analyst's role is one of making the model explicit, of using the model to obtain the elements of the responses, of enlightening the decision maker of the consequences of a certain type of behavior by translating them into terms that she understands, and perhaps of recommending (advocating, advising) one or a series of actions, or perhaps a methodology. The analyst's success will depend greatly on how he uses the resources allocated to him to construct the model, refine the problem formulation, verify the data, and choose an operational approach. The first and second parts of this book emphasize these features of decision aiding.

It is not enough that the analyst be convinced of his results; they are only useful if they assist in persuading the decision maker and, through her, the other stakeholders to conform to them. How deeply should the analyst become involved in the dynamics of the process? Would the complexity and technical sophistication of his work not render his arguments incomprehensible or, at least, suspect to the decision maker and other stakeholders? We shall return to these two questions several times in the remainder of this chapter, but we do not pretend to be able to answer them definitively.

2.2.4 Presence of a client

When the analyst and decision maker are different people, there is often little, if any, direct contact between them. A third actor then appears between these two – namely, the person requesting the study and responsible for allocating the means needed to conduct it. We call this person the **client**. A distinct client may also exist when there is direct contact between the decision maker and the analyst. Finally, the client may sometimes play the role of the analyst.

When the person requesting the study is neither the decision maker nor the analyst, this intermediary (client) is usually a high-placed person within the organization (cabinet secretary, vice president, head of a department, ...). She must have the power to manage the means necessary to conduct the study and have direct access to the decision maker, perhaps even being considered her representative. For example, in a problem concerning the choice of a highway location where the decision maker is a director of public works, the client may be a representative of the director, the leader of a design team, a representative of the environmental protection agency, the president of a citizens' organization, ...

Since she advises the decision maker and oversees the study, the analyst might wrongly think of the client as the decision maker. To illustrate the difference, consider the case of an individual responsible for approving loans. Suppose that this individual's immediate supervisor wishes to implement a systematic procedure that would leave only the most difficult requests to subjective assessment. Even though the supervisor decides on the implementation of the procedure, she is the client and not the decision maker: The model deals with the subordinate's approving or rejecting loan requests, and the procedure is conceived for her as the decision maker. Of course, in this role she must have views similar to those of the supervisor and, more generally, in accord with the policies of the general administration of her organization.

The client presents the problem to the analyst and usually oversees the work. She is usually willing to use her expertise, information, and contacts to help the analyst learn as much as possible of the class of phenomena and family of questions. In particular, she must ensure that the analyst does not attack a wrongly specified problem, i.e., one which is taken from its context or formulated in a way that will not allow a meaningful insertion into the decision process.

2.2.5 Aid and neutrality

The analyst cannot remain completely outside of the decision process if he wants his work to affect it. He is in some ways similar to a stakeholder of the second degree.[9] His role is to explain, to justify, to recommend, but he must do this independently of his own value system. It may be tempting for him to go beyond this role and become a stakeholder of the first degree, one who tries to change the problem according to his own value system and restrict the freedom of the decision maker.

In physics, and more generally in the case of phenomena relevant to the natural sciences, the model exists to reflect a generality (in appearance or in some underlying structure), and it can usually do so without altering or influencing future events. In the case of decision aiding, however, as soon as the behavior of a human being enters the picture, the model can no longer be considered to exhibit this type of neutrality. The modeling effort might, for example, cause a change in some value system, affect the development of preferences, or lead to the consideration of previously unsuspected possibilities.

Constructing or using the model will force the analyst to introduce what we shall call **voluntary hypotheses**. These are hypotheses that, by definition, could not be proven true or false, either because no conclusive tests could be designed or because they are imposed as policy. These hypotheses may concern the values of certain parameters (interest rates, growth rates, timing of a future event, ...) or the very structure of the model (consideration of certain scenarios, definition of decision variables, causal variables and data, existence and nature of a relationship, ...).

In the same spirit, the model cannot be considered neutral in the way that it omits or includes stakeholders other than the decision maker. Consider again the individual responsible for approving a loan. Suppose that she wishes to consider the general policies of her superiors toward loan approval, the information she receives from a questionnaire that accompanies the loan request, and her subordinates' perception of the person requesting the loan, a person with whom they are in direct contact. How will the analyst combine these elements to recommend approving the loan, not approving it, or requesting more information?

---

[9] *Even though this was not their intent, the works of the Centre de Gestion Scientifique de l'Ecole des Mines de Paris and the Centre de Recherche en Gestion de l'Ecole Polytechnique highlight this necessity. See, e.g., Berry (1983) and Riveline (1983).*

2.2.6 Aid and Objectivity[10]

If the analyst's work cannot always be neutral, can it lead to a truly objective model? We shall consider a model to be **objective** only if, in the eyes of a certain audience, it constitutes:

- an impartial and unbiased representation of the class of phenomena that it is to reflect within the context of the questions considered;
- an impartial and unbiased vehicle for investigation or communication, given the class of phenomena represented and the manner in which they have been taken out of their context.

This conception of objectivity implies that the analyst make clear and obtain acceptance of all of the essential hypotheses of the model, as well as the nature (voluntary hypothesis or not) of each. It also implies that the major factors of imprecision, uncertainty, and inaccurate determination that accompany the hypotheses are made explicit and that their influences are understood.

One can see how technical difficulties with perfect objectivity will arise from these two implications. Obtaining objectivity requires a good deal of effort, effort that the analyst may not wish to expend, especially if he is or becomes a stakeholder in the process.

Objectivity in decision aiding is faced with less technical difficulties, as well. These difficulties stem from the fact that the decision is made for and by a human being who, unlike a machine, possesses psychosensory and emotional sensibilities, as well as an aggregating and evocative intuition. As Trémolières (1975) writes: "reducing knowledge to objective knowledge, necessary in the physical sciences, has invaded current thought and swept away the other aspects of knowledge" (translated quotation). These other aspects of knowledge, even if they appear somewhat subjective, must be recognized in decision aiding if the process is to address human beings and not robots.

For this reason, decision aiding cannot dismiss certain features from the models on which it is based simply because the features stem from emotions or are difficult to quantify. In comparing different highway locations, for example, one must acknowledge features such as inequities in displacing or inflicting noise on a certain group of people, the destruction of historical sites, or the preservation of natural resources (see Bertier and de Montgolfier, 1971; Bétolaud and Février, 1973). This kind of difficulty appears in many other real-world problems, as well (see, e.g., Roy and Hugonnard, 1982; Roy *et al.*, 1986; Ostanello, 1990; Gregory *et al.*, 1993; D'Avignon and Sauvageau, 1996). Similarly, it is quite likely that the bias introduced when ignoring the uncertainty, imprecision, or inaccurate determination (see, Chs. 8 or 9) associated with numerical estimates will be stronger than that which would be present if these factors were acknowledged, but not captured precisely. Although these comments may seem obvious

---

[10] *The reader can also consult Ackoff (1977); GRETU (1980); Heurgon (1982); Hatchuel and Molet (1986).*

when stated explicitly, they are often implicitly contradicted in practice. For example, transportation analysts often ignore the difficulties of capturing features related to travel time savings of different alternatives by simply summing extremely small durations of time as if these are considered to be fully perceived and "saved" by each of a large number of users.

There are many models, then, that must incorporate elements which defy instrumental objectivity and physical logic if they are to be objective in the sense used at the beginning of this section. Many violations of objectivity result, therefore, from what we shall call **instrumental bias**. This bias is exhibited when the modeling effort uses instruments that produce seemingly objective or independent (of the observer) observations, even though the contents of the observations are less relevant than those requiring imperfect instruments producing estimates associated with a certain amount of subjectivity. This is somewhat like the person who, not being able to find her key after a night on the town, looks for it under all the street lights she passed on the way home, not because she has any reason to believe that she lost it near a street light, but simply because that is where she can see most clearly.

Other and perhaps more subtle biases exist as well. Wanting to get beyond a certain level of information or phase of the study confronted with complex, vague, and changing realities (like the preferences of a decision maker) can readily lead to a lack of objectivity. Restraining the field of imprecision or uncertainty – or of incomparability, in the case of preferences – often sets off a process of double selection (see, especially, Slovic, 1972). This process favors arguments that agree with preconceived ideas and adjustments that are based on excessive attention to extreme values and tendencies to overweight unlikely features.

So, taking what appears to be a cleaner, tighter direction may lead to a model that appears more reliable and more refined, but the model may well be based on prejudices and unrepresentative situations.

Finally, it is important to wonder how well an analyst, even one who is aware of these biases and concerned with objectivity, can detach his work from his own value system. Is it not inevitable, at least for certain classes of phenomena, that this value system will affect the model without his even knowing it (see, footnote 2 in Definition 2.1)?

In conclusion, it is intellectual honesty rather than objectivity that should be valued in decision aiding. This honesty is indispensable in making a correct deduction, as well as in obtaining a complete and unbiased picture of the situation. We wish to emphasize that these are the two elements which endow the analyst's work with a scientific character (see Piaget, 1973), since only they come into play when trying to convey the logical and experimental (suitable to reality) validity of the work to a broad community.

It is in this spirit that we have conceived of the proposed methodology (see Section 4.2). In particular, it is meant to assist model developers and users in distinguishing what is solidly established and, therefore, likely to attract the support of others, from what is debatable or arbitrary.

Chapter 3

REFERENCE EXAMPLES

SUMMARY

We introduce 12 reference examples to illustrate the notions discussed in the first two chapters. We shall continue to develop the examples throughout the remaining chapters of the book in an effort to clarify and highlight concepts as they are presented.

Throughout this book, we shall be using examples to define and illustrate the concepts, models, algorithms, procedures, and methodologies developed. We shall continue to refer to the simple family car, highway location, and loan approval examples begun in the previous chapters and introduce others when needed. We shall also develop 12 more complex examples gradually throughout the subsequent chapters.

In this chapter, we introduce these 12 examples. Each is based on an actual decision aiding case. The applications have been somewhat simplified and generalized, however, both to illustrate important methodological concepts and to avoid forcing the reader to enter into extraneous detail. We should point out that these examples are not intended to offer a step-by-step guide for professionals faced with similar problems. Rather, the objective is pedagogical. Finally, many of the numbers have been changed to protect the confidentiality of the data.

To help the reader follow the development of the examples more easily, we shall present an example number and short title each place the example is taken up in the text. We shall also specify the chapter and section where the example was last discussed. At the end of each presentation, we shall specify the chapter and section where the example will be continued.

3.1 INFRASTRUCTURE PROBLEMS

These examples might deal with a factory, a school, an airport, a rail line, or even a new production mode. As in the highway location example, the decision is generally considered irreversible and one-of-a-kind. It usually requires large expenditures and has important impacts on the lifestyle or environment of a segment of the population.

Example 1: Site Selection

In this example, a French engineering institute is currently spread among several different sites. The distances between the various sites and the condition of some of the buildings have made the physical facilities obsolete, especially when considering the progressive strategic plan that has been developed for the institute. It is, therefore, widely recognized that the physical facilities must be consolidated at a new site.

Factors such as the cost of relocation, the amount of time before the new site would be functional, and how the new site would fulfill the desires of the personnel, students, and graduates must all be considered. Another important factor is how the location can help fulfill the institute's new mission. In this mission, the institute is to:

- become a center of innovation and higher education;
- function as a catalyst to socioeconomic development and as an instrument for regional growth;
- gain an international reputation by exploiting its location in the Mediterranean basin.

This example is taken from a study (see Khouadja and Roy, 1975) requested by the director of the institute. In terms of the decision aid, the director is also the decision maker, as described in Section 2.2.2. The final decision is to be made by the regional Chamber of Commerce and Industry. This group would rely on studies conducted by the institute's administration and on its suggestions and those of the other actors who would be involved in the eventual decision process, a process that could be ill-defined at the outset. The actors can be described as:

- alumni groups and student associations;
- institute personnel;
- local authorities representing the region where the institute currently exists (and must remain);
- the organization coordinating land use at the national level;
- diverse organizations – such as the Ministry of Commerce and Arts and Crafts, Ministry of Industry and Research, Ministry of National Education – that might affect the process through their influence both on the Chamber of Commerce and Industry and on the institute.

(continued in Section 6.1.1)

Example 2: Commuter Rail Line

In this example, we consider a large urbanized area where policy makers have decided to improve the public transportation system. To increase access to different zones, especially for residents of a growing suburb $\mathcal{B}$, while using the existing infrastructure as much as possible, it has been decided to provide a rail link between $\mathcal{B}$ and employment zone $\mathcal{E}$. Providing such service does not present any major problems in terms of alignment. The difficulties arise, rather, from design issues related to number of stations and capacity-related features.

The interested parties have all agreed on five general objectives:

- minimize investment and operating costs;
- minimize access time to the stations and line haul times along the rail line;
- improve the compatibility among urban development, employment, and the transportation system;
- maximize the well-being of the transport users (increase comfort, safety, ...);
- avoid environmental disruption as much as possible.

(continued in Section 5.2.2)

3.2 NATIONAL OR REGIONAL DEVELOPMENT PROBLEMS

In these problems, the decisions possess an important political component. They cover a relatively long period of time and are usually subject to successive revisions. The difficulties are principally due to ignorance of or uncertainty surrounding the consequences of the possible decisions and to the sensitivity of compromise solutions to the value systems that come into play.

Example 3: Agricultural Development

Here, we consider the case of a developing country $\wp$ with a rapidly increasing population. This country imports a large percentage of its food for domestic consumption, and the imports are having serious effects on the country's balance of trade. Country $\wp$ could implement measures – such as forming rural zones, developing virgin land, irrigating dry areas, ... – that would lead to agricultural development and help maintain a high level of agricultural employment in $\wp$. Maintaining agricultural employment is a major objective of the public authorities, who are anxious to stop the rural exodus to the cities. Independence of the country's food supply has also been recognized as another major objective.

The issue is one of assisting the decision maker – in this case, the public authorities of $\wp$ – understand which agricultural plan would best suit $\wp$'s needs, so that necessary investments can be made and the measures required to adapt to changes in exports, imports, and employment can be undertaken.

(continued in Section 5.1.2)

Example 4: Water Resource Planning

Consider a region $\mathfrak{R}$ possessing a vast amount of underutilized hydrological resources. (Corsica is the region upon which this example is based.) Several technical studies have already been conducted to explore the possibilities of building dams at various sites, of diverting several rivers, and of implementing other large-scale projects. A general plan must now be chosen to facilitate farming, wine production, and tourism.

The different combinations of individual projects lead to many possible plans, and the number becomes even larger when the timing of the implementation is considered as a decision variable. The inhabitants of $\mathfrak{R}$ and tourists are actors in this process. In addition to these third parties, and acting on their behalf, are those actors involved in developing the general plan. These actors can be grouped under the following categories:

- IMIVAR (Institut de Mise de Valeur de la Région): This organization is particularly interested in the physical development and economic growth that would result from investments. It would like to increase the region's value as much as possible, particularly through technological growth. It is not too concerned with costs, especially long-term costs, and expects a large demand for services. Therefore, it wishes to supply ample capacity and does not worry about providing excessive capacity. It is concerned with the ease of acceptance by the other partners, however;

- the Ministry of Agriculture (of the state in which $\mathfrak{R}$ is found): The Ministry is responsible for financing investments. Like the IMIVAR, it is concerned with technological growth and the acceptance by other partners. Unlike the IMIVAR, however, it is not concerned with building to meet future demand, believing instead that demand can be satisfied as it develops. It is, therefore, against providing excessive capacity and is strongly in favor of decomposing strategies into parts that are as independent as possible. Finally, conservation of the environment and natural resources is one of its official objectives;

- the Jacobins: These are locally elected officials who believe in developing the regional economy in a fashion compatible with development at the national level. They consider the variety of activities served particularly important, as well as the ability to provide service as soon as possible. They are especially concerned with operating costs (and not very concerned with other costs), with ensuring adequate supplies of water, and with the attitudes of the other partners;

- the Girondins: These are elected officials with a much stronger desire for regional autonomy, and they oppose any developments that could inhibit it. They like strategies with independent components that can be controlled by local representatives. They are more concerned than the Jacobins with preserving the natural environment and with the number of individual economic units served, which they consider more important than the diversity of activities served.

The study should clarify the choice to be made, while analyzing its sensitivity to the value systems of the four different sets of actors.

(continued in Section 6.2.2)

3.3 ADVERTISING PROBLEMS

Whether they concern the contents of the advertisement, the selection of the media to use, the frequency of inserting advertisements in the chosen outlets, ..., advertising problems can lead to a myriad of intertwined difficulties. The successive decisions that help define an advertising campaign are, therefore, interdependent. The way in which the analyst defines the scope of what will be addressed will have an impact on the efficiency of his analysis and on the quality of the resulting aid he supplies. How he does so will be influenced by the generally repetitive nature of the decisions that he tries to clarify, as well as by the quantity and quality of data available.

Example 5: Media Planning

This example deals with only one (see Abgueguen, 1971) of many possible dimensions in advertising problems: selecting outlets for advertising. The objectives of an advertising campaign p have been defined, and it has been decided to use the written press media. The problem still remains of selecting the specific periodicals (magazines, newspapers, ...) in which to place advertisements.

Advertising agency $\mathcal{A}$ is continually faced with this problem. The department that selects the periodicals proceeds somewhat as follows.

a) Choose several (a dozen, at most) pertinent points of view to determine the main advantages that each periodical offers in light of the objectives of the campaign plan;
b) Screen the numerous foreign, national, regional, and local periodicals with which $\mathcal{A}$ works to determine a dozen, or so, that should be considered further;
c) Determine the correspondence between each periodical retained in b) and each point of view chosen in a);
d) Discuss (especially with the client) the periodicals that seem most promising according to the results of c) and construct two or three combinations that seem to be effective;
e) Investigate in more detail the consistency of the combinations formed in d) (primarily as a function of duplication and complementarity of the periodicals in a plan) and decide if one of them is satisfactory; if not, return to d).

Although the director of $\mathcal{A}$ feels that Step e) is too complex to be modeled formally, she does believe that a more systematic approach to the previous four steps would lead to better results in Step e). Therefore, she requests a study that would offer the following improvements:

- at a): A survey of points of view, and for each, a survey of the characteristics of the plan that would lead to choosing or rejecting the point of view;
- at b): A less drastic screening process that retains several dozen, instead of only one dozen, periodicals;
- at c): A codification of the correspondences between points of view and periodicals;
- at d): A flexible procedure that could guide the discussion on the combinations presently considered and stimulate thinking so as to arrive at other effective combinations.

(continued in Section 6.1.3)

3.4 RESEARCH AND DEVELOPMENT PROBLEMS

All types of organizations must decide whether to support or reject, prioritize or abandon a given research or development operation. These problems might involve periodic (see Example 6) or once only (see Example 7) decisions. Modeling these decisions is almost always complicated by the vagueness associated with the contents and progress of these operations, by the difficulty in quantifying the many consequences that will determine their success, and by the hypothetical nature of this success and of the investments that will be required.

Example 6: Research Projects

Here, we consider an industrial organization *O* whose Research and Development Division accepts and rejects funding requests for research projects submitted from the rest of the organization. The Division is given a budget by the upper management, and the level of this budget changes from year to year.

The decision maker in this case is a committee that determines the general research agenda. This agenda is implemented at the research project level, where each project has well identified objectives and potential benefits to *O*. The research itself can last from one to five years and might require the use of several research teams in succession. Thus, the funding decision is made at the finest level, that of the research task. A research task is that part of the research project for which there exists a sub-objective assigned to a single principal investigator and whose expected duration is 18 months or less. A project can consist of several tasks or of one single task.

The different divisions of *O* submit to the committee:

- either proposals for new projects, describing in detail the initial tasks;

– or proposals for continuation projects, describing in detail the tasks that are to follow immediately and results from previously completed tasks.

These proposals can be submitted at any time, but most are submitted during the last quarter of the year.

The committee must make a funding decision on each task before the end of the year in which the proposal is received. It cannot fund tasks proposed for later times. When a proposal is received early enough in the year, the committee can decide either to fund it immediately from the current year's budget, to reject the proposal immediately, or to postpone the decision until the end of the year and consider it in the following year's set of projects.

(continued in Section 5.1.2)

Example 7: Industrial Development

This example treats the choice of technical and commercial actions to be made by a large company *C* that has a monopoly on the distribution of electricity (see Charpentier and Jacquet-Lagrèze, 1976). The public regulatory agencies have agreed to allow *C* to promote new industrial uses of electricity in the next few years.

Assume that a very complete data base has been developed covering:

- the different industrial sectors (metallurgy, electrical industries, mechanics and electronics, construction material fabricators, chemistry, agricultural and food industries, ...) and the production methods that they are or could be presently using;
- the possible applications of electrical energy (drying, heating, mechanical movements, ...), as well as the various technical means of providing electrical energy for each application (electrical resistance, radiation, induction, pressure, ...).

The problem is how to use this data base to build a development **strategy** and, more specifically, how to aid in choosing actions that should be undertaken in the short term.

(continued in Section 5.1.1)

3.5 OPERATIONS PROBLEMS

The decisions considered under this title are those that are regularly made in a company dealing with production or services or in an administration where the goal is to function as best as possible. Therefore, the decisions are fairly reversible but not always routine, as in Example 8. Improving the quality of products and services, or simplifying management and making it more reliable, will usually be more important than any direct benefits that might be achieved.

Example 8: Airport Operations

Let $\mathcal{L}$ denote an international airport where the increased traffic, changing requirements of the passengers, and improved technologies have resulted in a need to upgrade the terminals. Past interviews, surveys, and preliminary technical analyses have led to a somewhat precise description of fragmented actions addressing specific aspects of the problem; the different fragmented actions can be combined to develop more comprehensive actions. The number of these fragmented actions (around one hundred) makes it impossible to conduct a detailed study of each.

The authorities of $\mathcal{L}$ would, therefore, like to develop a general feel for the various possibilities and to screen them to obtain a short list. For this reason it has requested that the following aspects be considered:

a) time savings for the passengers;
b) reactions of the passengers;
c) impact on peak periods;
d) investment and operating costs;
e) complexity of the necessary negotiations (e.g., legal restrictions, employee acceptance, industrial relations);
f) time for implementation;
g) time required to finish a detailed analysis.

(continued in Section 6.2.2)

Example 9: Engine Assignment

In this example, a dispatcher must make routine decisions in relatively short periods of time. The decisions involve which engine (locomotive, tank truck, ...) is to be assigned to which task (powering a train, transporting liquid products, ...). The fleet consists of approximately 30 similar, but not identical engines.

At any moment, most of the engines are performing tasks that must be completed before they can become available for reassignment to a new task. The time when any engine will complete its task is not known with precision until the task has actually been completed. A free engine can be immediately assigned to a task. The time between the assignment and the actual beginning of the task is an important consideration in the problem, however.

Each task i is specified by:

- its requested starting time t_i: the dispatcher must adhere to this time as best as possible;
- the required state E_i of the engine that will be assigned to it: this state captures characteristics of the engine type (e.g., is it powerful enough for the task?) and of its present condition, which will depend on the last task performed (e.g., does it need to be cleaned before executing the task? where is it currently located?);
- the anticipated execution time of the task d_i: this time is independent of the specific engine performing the task, but the actual execution time might (but only rarely) be significantly greater than d_i; i.e., the engine performing task i might not be available until after $t_i + d_i$;
- the state E_i^* of the assigned engine when it becomes available: the components of E_i^* that depend on i will be things such as where the engine is when it becomes available, whether the engine needs cleaning or not, ...

As for the dispatcher, she has access to:

- a terminal (connected to a computer) which can be used for calculations – such as the time $D_k(E_q^*, E_i)$ required for an engine k, which was in state E_q^* when completing its last task q, to enter into state E_i required by task i; we add here that the most important costs to consider in this problem are those associated with the nonproductive time in which the engine passes from one state into another (e.g., deadheading of a locomotive; cleaning of a tank), as opposed to direct costs (e.g., fuel consumption), which can be ignored;
- a telephone linking her to a superior who can provide additional information when needed; this information relates to deviations from the normal plan, such as beginning task i after t_i or assigning an engine whose actual condition does not exactly place it in state E_i (which increases the likelihood of either not being able to accomplish the task or of requiring additional costs to do so);
- a telecommunication system keeping her informed of actual assignments and of the status of available engines and new tasks.

In addition:

- the time for an engine to begin the next task after having completed a preceding one can be as large as the time to execute the task itself (which is on the order of several hours);
- at time t only those tasks with requested starting times after t + 60 (one minute being the time unit) can ordinarily be considered for assignment or for having their characteristics modified by the superior;
- the number of tasks not already being executed that require an engine at any time varies between 20 and 50.

Under these conditions, how can the dispatcher be helped to "best decide"?

(continued in Section 5.2.2)

3.6 SELECTION PROBLEMS

Here we are interested in selecting *p* (a fixed number to within a few units) individuals from a population *P* (a number much greater than *p*) of candidates applying for a job, for admission to a school, ... One fairly traditional procedure for decision aiding is that of ranking individuals according to some type of test. Although this procedure has certain advantages, we note that it can also be subject to what we have termed an instrumental bias in Section 2.2.6. Moreover, the amount of effort spent on evaluating the different candidates in this traditional approach is usually the same, regardless of whether the candidates are only marginally qualified for the position they are seeking or whether they obviously warrant immediate acceptance or rejection. In in the marginal cases, the director of personnel or the admissions committee may also need or wish to base decisions on other information and opinions.

Example 10: Application Packages

Like Example 1, this example comes from an academic setting, but the connection between the examples stops there. The crux of the ulterior developments could equally well apply to other settings, such as hiring groups of individuals into large companies.

In this example an academic unit wants to improve its admissions procedures by developing a method that would assist in evaluating candidates based on formal application packages and, in some cases, on complementary interviews. Interviews would be required of those applicants whose formal packages – consisting of approximately 20 elements (see Moscarola and Roy, 1977), such as marital status, previous grades, results of psychological tests – would not lead to a clear admission or rejection. Since the number *p* of students forming a class is fixed to within a few students, the method would have to insure that the number of candidates admitted on the basis of their application packages alone would not exceed or even get too close to *p*.

This new procedure would come under the authority of a committee consisting of five or six members. The committee is interested in an automatic screening of the application packages, so as to decide whether to immediately accept or reject each applicant, or to require that he or she undergo complementary interviews and be evaluated by less automatic methods. (The committee would always be able to override the "automatic decisions," however.) This automatic preliminary procedure would not only reduce the work load of the committee (the number of applicants is very large), but might also reduce certain undesirable phenomena such as tendencies to favor certain criteria, the effects of members' moods on different days, ...

(continued in Section 6.1.2)

3.7 MANUFACTURING PROBLEMS

What manufacturing conditions should be specified to guarantee a sufficient level of quality? What can be done to meet these conditions at minimal cost? How can the workforce be organized so as to satisfy the requirements and aspirations of the personnel, safety organizations, ...? These questions are among those that have contributed to the rapid expansion in the use of linear programming (most notably in the petroleum industry). The first example treated under this title is illustrative of the classical manufacturing problem, while the second one is quite different.

Example 11: Product Composition

Consider a product such as rubber (but one could also consider detergents, paint, perfume, ...), whose production requires multiple components. To make 100 sheets of rubber, one adds specified quantities of elements such as stearic acid, sulfur, pine tar, and zinc oxide. Different specifications lead to different properties, such as external heat resistance, fracture stress, heat build-up, and fatigue resistance. Increasingly, diversified uses of rubber have generated constantly changing performance requirements.

Because of the interactions among the elements making up a given product, the performance level of each property is a complicated function of the quantities of the elements in the composition. We shall assume that the technical division of some manufacturer $\mathcal{F}$ knows the analytical form (e.g., linear or quadratic) and has estimated the parameters that would provide good approximations to each performance function. These approximations are valid only within fairly narrow ranges of the component quantities – i.e., certain coefficients would have to be adjusted if the quantities fell outside these ranges.

Confronted with constantly changing performance specifications and an increasing variety of products, the technical division finds it increasingly difficult to determine the component mixes that would "best" satisfy the needs and wishes of the company's clients. The upper management of $\mathcal{F}$ has, therefore, asked a consulting company to develop a systematic decision aiding methodology to replace its traditional, relatively empirical approach. The technical division expects to use this methodology to determine the quantities of each product component that would either minimize cost or maximize certain performance characteristics when a new request is made.

(continued in Section 5.2.2)

Example 12: Plant Organization

In this example, we consider an automobile assembly plant (see Giordano and Suquet, 1976). The diversity in organizational possibilities of assembly plants is much greater now than it has been in the past, when one only had to consider dividing tasks and balancing production lines.

Suppose that the plant to be considered must be completely reorganized and that many modern organizational strategies are being considered. Management is primarily interested in decreasing the psychological stress of routine tasks on the workers. It feels that a complete restructuring of the assembly plant $\mathcal{M}$ is warranted. More generally, the plant is to be considered a system with interactions among various aspects – e.g., technical aspects (logical execution, compatibility with worker capabilities, ...), economic aspects (daily production, rejection rates, ...), human aspects (physiological characteristics, ...), and social aspects (absenteeism, strikes, ...).

The new organization might, for example, combine assembly lines (which could use long or short cycles), assembly in small groups (which could be autonomous and self-managed to varying degrees), and even individual assembly. A new organization would be defined by the various tasks and how they fit in the plant, while specifying the product transfer mode in assembly, the major cycle times, the level of various inventories, and the like.

Besides the analyst, the principal actors who would be involved in the decision problem are:

- the supervisor of plant $\mathcal{M}$;
- the company safety officer;
- a representative of general management who can authorize expenditures and specify certain constraints;
- labor representatives;
- third parties, who in this case are the workers.

We also assume that this is the first time that the company has tried to reorganize a plant in this way and there is, therefore, little information available to the analyst. The impacts of various types of organizational structures are not well known in the company, and comparing two structures would pose difficulties. The only data that the analyst can call upon are those dealing with existing plants, either plants within the company or plants for which there is information available in the professional literature.

Under these conditions, how can the analyst aid in choosing a new organizational structure for plant $\mathcal{M}$?

(continued in Section 6.2.2)

Chapter 4

PHASES AND OPTIONS OF AN APPROACH TO DECISION AIDING (GENERAL IDEAS OF THE METHODOLOGY)

SUMMARY

In Section 4.1, we introduce the basic ideas of value system, informational system, and relational network; define the term actor; and present the interrelated concepts of study phase and decision process development state (PDS).

In Section 4.2, we sketch out the major options of an approach designed to assist the analyst in recommending or simply participating in the process, while acknowledging the decision maker's ultimate freedom. These options serve as a guide for the proposed methodology and the organization of the remainder of the book.

4.1 NOTIONS OF STRUCTURE

4.1.1 Preliminaries[1]

Like all general methodologies, ours requires basic concepts that are usually difficult to capture precisely but play an important role in structuring the approach. Rather than defining these concepts rigorously, we try in this section to give a feeling for their implications and how they can help focus the discussion.

The first of these notions is that of a **system**, a term that we have already used. A system commonly[2] refers to **a complex entity considered (with respect to objectives) as an organized unit that retains its identity during certain evolutions and consists of elements and relations among these elements that are defined and differentiated by their contribution to the unit**.

By **value system** we mean the somewhat implicit system that underpins the very basis of the value judgments of an individual or a group. These judgments can be relative (better, worse, ...) or absolute (good, bad, ...). The value system will greatly influence what eventually becomes important, as well as the **objectives** and **norms** that are formed, although the objectives and norms are often considered to come first in order to justify or to provide a hierarchy for the value judgments and the tangible behavior to

---

[1] *For further development of the ideas presented in this subsection, we refer the reader to Jacquet-Lagrèze (1981).*

[2] *See especially Morin (1977), pp. 100 and following, and Le Moigne (1977).*

which they lead[3]. This is why one sometimes speaks of a system of objectives and a system of norms. Any information that individuals consider and use during the decision process can only be considered and used as a function of their access to what we call an **informational system**.

Finally, there exists a somewhat solid framework of influences, alliances, coalitions, pressures, ... between a given individual and all the others involved in a decision process. This framework, which we call a **relational network**, is strongly influenced by the value system and by the informational system; it also affects how they are connected.

We can now better define what we mean by an actor.

DEFINITION 4.1.1: An individual or a group of individuals is an ***actor*** *in a decision process if she directly or indirectly influences the decision by her value system. This influence can be a first degree influence, resulting from the actor's intentions, or a second degree influence, resulting from the way in which she influences other individuals to intervene. Moreover, for a group of individuals (entity or community) to be considered as a single actor, no distinction should exist in the value systems, informational systems, and relational networks of the different members of the group.*[4]

The examples in Chapter 3 (especially, Examples 1, 4, 7, 12) provide a more concrete interpretation of this definition.

In this book, then, the term "actor" denotes a social subject in the sense given by Boudon (1977, p. 12) and Crozier and Friedberg (1977) when he refers to a homo-sociologicus. We emphasize that our use, "does not imply the image of a rational homo-sociologicus, but of an intentional homo-sociologicus ... conceived as being driven by the objectives he wishes to attain and by the way in which he represents the means that allow him to attain his objectives" (translated quotation).

An actor's influence on the decision could result from:

- an intentional action undertaken to affect the course of the process directly so that her own preferences will prevail; in Chapter 1, we called such an actor a **stakeholder**;
- her preferences being considered by others, even though she plays a passive role in the decision; in Chapter 1, we called such an actor a **third party**;
- an effort that takes into consideration the different value systems involved and simply tries to shape the course of the process so that it conforms to certain intentions; in this

[3] *See, for example, the part of Example 4 in Section 3.2 related to the IMIVAR, the Ministry of Agriculture, the Jacobins, and the Girondins.*

[4] *This in no way implies that these systems and networks are identical, but only that the analyst considers it convenient and realistic to reason without distinguishing among them.*

case, the actor is the **client**, the **analyst** (see Section 2.2), or, perhaps, a consultant or mediator.

As Hirsch, *et al.* (1977) emphasize, an actor can only be specified in the context of a given process. It is, therefore, difficult to define the actor when several processes are occurring at the same time.

Consider again the distinction between entity and community introduced at the beginning of Chapter 1. In both cases, the actors are identified as groups containing more than a single individual. The expression entity is used when the different individuals comprising a group are specified in such a way that there is no ambiguity in deciding who belongs to the group. Community, on the other hand, is used to define a family with imprecise relationships that somehow group together individuals who are not well specified, but who cannot be differentiated either by their value systems, their informational systems, or by their relational networks.

4.1.2 Study phase and decision process development state

Resolving an intermediate problem, changing the attitude of a stakeholder, transferring the responsibility for a decision to a higher or lower level in an organization, altering what is considered possible by a partial decision, ..., are all the types of circumstances that appear as benchmarks in the evolution of a decision process. We shall speak of the "**process development state**" (PDS) to signify the situation created by a specific critical point that launches a **study phase**. In such a situation, the relevant characteristics will not change until a later critical point occurs that impacts the final decision (see Chapter 1). The impact on the final decision results from a modification of the conditions under which the decision process unfolds or the analysis is conducted.

*DEFINITION 4.1.2: The "**process development state**" (PDS) is the set of facts and working conditions that characterize the situation created by the evolution of the process. These facts and conditions summarize the previous history in data and constraints (resulting from partial decisions and preliminary hypotheses) that will affect the way in which the analysis is conducted and will usually be taken as fixed from that point on.*

By definition, the degree of the analyst's accessibility to data remains the same during a PDS. This does not, however, preclude that he may acquire information and that the situation may evolve. Similarly, unless the analysis eventually shows them to be poorly identified, during a PDS there is no substantial change in the roles of the various stakeholders, in the attitudes of the different actors toward the problem, or in what is believed to be the object of the decision. The analyst's contribution may very well be one of making access to some data possible, transforming a stakeholder's attitude or decision maker's preferences, shifting the level where a decision is taken, or highlighting the advantage of making intermediate decisions. If such a contribution renders the analyst's present model inappropriate for the rest of the study, there is a change in the PDS. If, however, the model remains useful for the following elements of the study, we do not consider there to have been a change in PDS.

The PDS is the indispensable framework for fixing the given phase of the analysis. The constant conditions characterizing it lead to useful deductions. They also specify the nature of the next critical point of the analysis and situate it with respect to the comprehensive decision. They will, therefore, guide the analyst's subsequent activities.

The fixed characteristics that describe a PDS can lead to several phases, either because they lead to independent critical points or because they lead to different conceptions of a single critical point. Like the decision process, the analytical process is not necessarily linear. More precisely we will say that, by definition, **there is a change in analytical phase when there is a change in PDS or a change in model.**

4.2 THE PROPOSED METHODOLOGY

Consider a (not necessarily well-formulated) problem concerning a decision (or a bundle of decisions) and a certain (not necessarily well-defined) PDS confronting an analyst (see Chapter 3, Examples 1 to 12). With the help of the client, this analyst explores the problem, gathering information on preceding studies and on previous critical points in the decision problem. He identifies the decision maker and describes the PDS that will form the framework for the phase that he will undertake. When appropriate, he considers how this phase will connect with later phases or phases that are occurring in parallel. Even at this preliminary level, commitments must generally be made to basic options that tend to "frame" the analysis. These options are primarily related to characterizing the PDS and the nature of the envisioned critical points. The expected critical points will influence the direction of the current phase and how the future of the process will be divided into individual phases. We shall not deal with these options here.

The options with which we shall be concerned in this volume are of a more technical nature. They deal with analyzing and modeling or with the procedures used to acquire information and obtain solutions. We divide them into four levels.

This volume is primarily concerned with the options taken at the first three levels. We, therefore, divide the rest of the book into three parts so that the presentation of the methodology corresponds as closely as possible to these levels I, II, and III. We touch upon the options of Level IV in the third part but leave the theoretical developments, detailed presentation of methods, and illustrative case studies to another volume. In this following work, we shall also go more deeply into the options of Level IV and clarify others found at the different levels.

4.2.1 Level I: Object of the decision and spirit of recommendation or participation

In light of later critical points – specifically, that which turns the present PDS into the following one, which could be the final contribution to the comprehensive decision – how should the decision be modeled? How can the various actions be differentiated? What will determine which actions are considered possible? The analyst takes a certain number of options when confronted with these questions. Analyzing these options will lead to the concepts of action and potential action in Chapter 5.

At the same time, the analyst takes a position on the spirit in which he sees his participation, indeed his **recommendation**. We use the word "recommendation" to emphasize that both the analyst and the decision maker understand that the decision maker remains free to act however she wishes, even after the recommendation is made. In this way, the terminology is different than that used in much of Multi-Criteria Decision Aiding and that used in the original, French version of this book, where the term "prescription" is used in the medical sense of the word to signify "stating a suggested treatment" (translated quotation based on Robert, 1968). We prefer to use the more general term "recommendation" to the term "prescription," however, since for many Anglo-Saxon authors (see, e.g., Watson, 1991), prescribing implies a stronger sense of aiding someone approach an ideal state, a concept that we believe to be unfounded in decision aiding. (For more details on this distinction, see Roy, 1993.) Rather, the analyst only states a simple advocacy of a specific behavior that has been thought through as scientifically as possible. To determine his participation or recommendation, the analyst is, therefore, led to place the present phase of study with respect to the poles formed by the four reference problematics presented in Chapter 6.

4.2.2 Level II: Analyzing consequences and developing criteria

To what extent will the modeling of the decision influence the evolution of the process? What consequences of the possible decision could be relevant to the objectives and to the value systems of the stakeholders? Which of these consequences must be explicitly modeled, and how? How helpful will each be in clarifying the decision, given the factors of imprecision, uncertainty, and inaccurate determination in the process? How can criteria be constructed that recognize these consequences and factors? The second level discusses these questions.

Chapter 7 considers the fundamental concepts and the theoretical bases required to understand and develop any model reflecting preferences. Although these concepts and developments are essential to Level II, they are not limited to this level. Therefore, they are considered somewhat separately in the second part. Chapter 8 then examines the structuring of a preference model to analyze the multiple consequences of the possible decisions.

The evaluation model, which conveys the results of the consequence analysis, is generally too complex to be used directly in decision aiding. Rather, one or several criteria must be developed. Chapter 9 covers this concept, as well as common ways to proceed from evaluating an action on all or some of the consequences to determining a performance measure of the action along a criterion that synthesizes the relevant consequences.

At this second level, the analyst must above all construct appropriate criteria and analyze their potential to form the basis for insightful comparisons among the actions.

4.2.3 Level III: Modeling comprehensive preferences and operationally aggregating performances

Options at the previous two levels can generally be taken independently of the decision maker's, or any other actor's, value system. It is much more difficult to do so with options at Level III, however, where the options bring about two types of concerns that are difficult to separate.

Given the usually large range of possibilities for defining the criteria, how should one (single criterion analysis) or several (multicriteria analysis) be selected to "best" capture the essence of the consequences for decision aiding? What is required if a certain family of criteria (which could, perhaps, be a single criterion) is to play its role in the analysis and generate an acceptable dialogue among the various stakeholders considered by the decision maker? Should each of the criteria in the family be considered as an instrument that will describe the actors' intangible preferences or, on the contrary, that will help unestablished preferences emerge, evolve, and perhaps converge? Chapter 10 deals with these concerns.

In multicriteria analysis, how can an action's performance measures on the various criteria be aggregated so that the action can be said to be good or bad, better or worse than another? What information – for example, related to the relative importance of the criteria – should be used and how? In Chapter 11, answering such questions will lead us to present the important notion of an operational approach. There, we present three very different approaches that cover the majority of those found in practice. The option chosen will greatly affect how the decision aiding effort fits in the overall decision process and what information will be collected and developed for modeling or determining comprehensive[5] preferences. The options taken here will largely determine the procedure that leads to the recommendation.

4.2.4 Level IV: Investigating and developing the recommendation

The decision maker's questions might lead to elements of response that make the resulting recommendation obvious. When this is not the case, we arrive at Level IV with a need for formal procedures designed to acquire and process information that lead to "solutions" to specific problems. The procedures available to the analyst will depend on the problematic and the operational approach that he chooses. As a function of these, we can distinguish among:

- selection procedures based on modeling comprehensive preferences and developed to lead to a choice;
- assignment procedures based on modeling comprehensive preferences and developed to lead to a sorting;

---

[5] *That is, preferences that consider all the criteria, as opposed to a preference restricted to one criterion, with all other things considered equal.*

- ordering procedures based on modeling comprehensive preferences and developed to lead to a ranking.

The analyst's primary difficulty when faced with options at this level result from a lack of time or a lack of useful information. This makes the decision maker or the other actors unable to understand the logic and mechanics of the procedures (tools) that will be used. In other fields, one is usually ready to trust the technician or the specialist. In decision aiding, however, the analyst rarely enjoys such trust. The nature of the stakes involved and the diversity of the actors' representation and logic systems make the debates that can arise at this level more difficult. Even so, the quality and the impact of the recommendation depend greatly on the analyst's ability to overcome such difficulties and fit into the decision process and, in some cases, to involve the eventual users of the tools in their development.

4.2.5 Comments

These four levels of options should not be considered to occur in series, where one stage cannot begin until the preceding stage is completed. Some of the options at Level I or II may not be taken until reflecting on the options at Level III. Preliminary results at Level IV may cause the Level I options to be reconsidered.

Similarly, the analyst's recommendation cannot be reduced to an isolated act that marks the end of a more or less complex path (e.g., I, II, III, II, IV, I, II, IV) through these levels. Indeed, the elements (synthesized information, answers to questions, suggestions, ...) of the conclusions that close a certain phase of the study are all essential components of this recommendation. The analyst will usually not be able to remain outside of the process along the way. When taking certain critical options, he will not be able to remain an observer, exerting no influence; nor should he try to do so! The reflections caused by his questions, the paths that he decides to follow in conjunction with the stakeholders, the data he gathers, the preliminary answers he furnishes, the intermediate results he communicates, the propositions he makes during the analysis, ..., all contribute to clarifying the decision. As such, all these acts are a part of the recommendation, in the sense that the term is used in this book.

Although primarily motivated by technical considerations, these acts will usually be critical to the final impact of the analysis. They are what make the analyst the stakeholder of the second degree discussed in Section 2.2 or what, on the contrary, lead him to reject this role. For this reason, decision aiding implies a minimum amount of insertion in the decision process: Decision aiding is not done "for" but "with" the actors of the process. Relations with third parties can be especially crucial (see Emsellem, 1976).

Finally, let us note that the analyst will face other, nontechnical options, which we shall only consider indirectly in this book. These concern the types of arguments and the means of communicating that the analyst or the client uses in the course of the analysis. Whether these are meant to facilitate understanding or to gain acceptance for the analysis and the successive options, to convey results or to build a consensus on certain doubts or beliefs, ..., many choices can be made, and these, too, can often be important to the quality of the analyst's insertion in the decision process[6]. Insufficient attention to these options will either lead the analyst to address the wrong problem or make him unable to get others to understand and therefore accept his results. These latter options may be related to and even conflict with the former, more technical and scientific ones. In general, when the latter options influence the former ones too greatly, the analysis loses its desired role, either because its scientific character is degraded and its nature becomes partisan or because the analyst becomes a stakeholder of the first degree, in other words, a participant with a vested interest.

[6] *These considerations are illustrated in concrete cases in Jacquet-Lagrèze, et al. (1978), Jacquet-Lagrèze and Marchet (1978), Major and Moscarola (1979).*

LEVEL I

HOW TO DETERMINE WHAT IS POSSIBLE AND IN WHAT TERMS TO FORMULATE A PROBLEM

Chapter 5

ACTIONS AND DECISION AIDING

SUMMARY

In Section 5.1.1, we define the term action. We distinguish between actual and dummy actions and realistic and unrealistic actions and illustrate these concepts by continuing the Industrial Development example. In Section 5.1.2, we distinguish between a comprehensive action, the execution of which excludes the execution of any other action introduced in the model, and a fragmented action, which can be combined with other actions for joint execution in the framework of the final decision. We illustrate the distinction in the continuations of the Agricultural Development and Research Project examples.

In Section 5.2, we define a potential action as an action that is temporarily assumed to be possible for the decision aid. We also present the idea of the set of potential actions on which the decision aiding effort is based during a phase of the study, which we denote by A. We present the conditions of internal and external stability and define A as stable (fixed and permanent) when these conditions are fulfilled, and A as evolving (revisable or transitory) when they are not. The concepts of Section 5.2 are illustrated through continuations of the the Commuter Rail Line, Product Composition, and Engine Assignment examples.

5.1 THE CONCEPT OF AN ACTION

Depending on the problem and the decision process development state (PDS) considered (Def. 4.1.2), the application point of the decision aiding effort can be a site selection (Ex. 1), a project or a variant (Ex. 10), a rate or a formula (Ex. 11), or a somewhat more complex configuration (Exs. 7, 9, 12, ...). We provide a general characterization of this application in the first subsection. In the second subsection, we consider different conceptions of the decision aiding effort.

5.1.1 Definition and examples

What the decision maker expects from the decision aid (Def. 2.2) will evidently be influenced by the different actions she considers to be related to the critical point or points that would follow the current phase of the study. Which location should be chosen for the projected highway (Section 2.1.1)? Which model of car should the family purchase (Chapter 1)? What should be the fate of each of the loan applications (Section 2.2.4)? What fraction of a total should be allotted to each group requesting certain subsidies? These simple examples show that the type of action forming the application point of the decision aid can be a certain physical object or variety (a variant of the location, a model of car), an element or support to be acted upon (a given loan application that can be accepted, rejected, or put aside for the time being), or a state that will take on certain characteristics of the decision (numerical values describing the subsidy allocation). In spite of this diversity of possible forms, we can introduce the general concept of **action**.

*DEFINITION 5.1.1: An **action** "a" is the representation of a possible contribution to the comprehensive decision that can be considered autonomously with respect to the decision process development state and that can serve as an application point for the decision aid. The application point is, then, sufficient to characterize "a".*

We should clarify our use of the term "autonomous," as well as add a few remarks, before illustrating the definition through a reference example.

Every action "a" introduced in a model must have meaning by itself. To say that it represents an autonomous contribution to the final decision means that it can be isolated from all the other actions without losing its decisional impact or its value as an application point for the decision aid. Thus, when considering different geographical variants of the location of a highway segment, design options such as constructing a tunnel, building below grade, or building at grade, are not independent enough in themselves to define actions. On the other hand, if the project requires that all the possible locations coincide at some critical site, each design option at this site might be considered an action.

Our definition does not incorporate any idea of feasibility or realism in the concept of action. Any division of subsidies in the above example, even the most unreasonable, can define an action. For clarity in decision aiding, it proves useful to make the following distinctions:

- **Actual** actions stem from a completely developed project that can be executed. On the other hand, **dummy** actions correspond to an idealized project, or one that is incompletely developed or even hypothetical.
- **Realistic** actions correspond to a project whose implementation could be reasonably foreseen. On the other hand, **unrealistic** actions are those that might satisfy incompatible objectives, but which provide a good foundation for discussion and reasoning.

A dummy action may be realistic or unrealistic. In Chapter 9 we shall refer to a particular category of dummy actions (realistic or not), which we call **ideal** actions. By this, we shall mean actions that correspond completely to the descriptions that one gives to their consequences.

We note that the options which some decision maker can take to manage the decision process or to change the conditions under which it unfolds will not constitute actions, except in those very special cases where the application point of the decision aiding effort is one that explicitly recognizes them as such. The study of such options (see the end of Section 4.2) requires a more complex methodology than that proposed in this book (see, especially, Lesourne, 1977; Norese and Ostanello, 1989 ; Ostanello, 1990).

We emphasize again that the nature and characteristics of the actions depend on the present process development state. Coming back to the problem of the subsidy allocation, an action a might be defined in terms of the amount of subsidy to be allocated to each party. Or, as a first cut, it could be defined at the level of each beneficiary, in terms of whether his or her allocation would increase or decrease with respect to the preceding year. This latter definition would lead to a completely different definition of the action than that considered in the example up to now (see Moscarola and Roy, 1977). We shall return to this feature in Section 5.1.2.

Finally, according to Definition 5.1.1, each automobile model or each loan application can lead to an action that, at the risk of a certain stretching of the language, will be identified by its "support," i.e., by the specific model of automobile or loan application. The rest of this chapter will show that thinking about the action "model of automobile a_f (proposed by the daughter in Chapter 1)" or "loan application a_m (submitted by Ms. m)" as a support of a question (should we look into model a_f? should we accept, reject, or hold off on application a_m?) will allow us to get to the heart of the problem in an operational manner. By characterizing each action by its application point in the framework of the decision aiding effort, we can define the action without having to consider whether the analyst will eventually optimize, sort, rank, or simply evaluate during the current phase of the study (see Chapter 6).

Example 7: Industrial Development (from Section 3.4)

The process development state can be briefly characterized as follows:

- in agreement with the public agencies, it has been decided that *C* should promote new industrial uses of electricity;
- the decision that currently needs to be clarified is what the marketing division of *C* (the client) should promote; the choice will ultimately rest with the Board of Directors of *C* (decision maker);
- among other things, the analyst has access to a data base on the set of industrial sectors and the set of applications for electric energy;
- given the technical and commercial nature of the study, several departments of *C* have been asked to participate in arriving at this decision and to formulate their priorities;
- the critical point that would logically indicate the end of this stage of the study and the beginning of the next is a meeting of the Board of Directors, which will take place within four to six months.

Under these conditions, the best way to get a handle on the actions is to define them as **sector-application** couples. By sector-application couple, we mean the combination of a given industrial sector with an application. An application is taken to mean an industrial operation that could be achieved in a specific manner. For example, drying by infra-red and drying with heat pumps would represent two different applications.

In this case, the "product" of what we have called the comprehensive decision is the entire development strategy for new industrial uses of electric energy. Those sector-application couples that the directors will declare as warranting near term promotional efforts will contribute to this comprehensive decision. The sector-application couple can be considered an application point in the decision aid, in the sense that it uses information that is available and that it addresses the critical point of the process considered.

Several hundred couples could be constructed from the existing lists of sectors and applications. Although each would be self-contained and would, therefore, represent an action, some of the couples could be immediately rejected as not being viable candidates. In the rest of this example, we shall assume that various working groups have screened the potential couples and have retained 124 of them as potentially warranting promotional priority. We shall call A the set of these 124 actions.

(continued in Section 6.1.2)

5.1.2 Comprehensive and fragmented conceptions: identification problems

Considering the purchase of an automobile, the location of a highway segment, or the determination of a vector of subsidies can lead to defining actions in such a way that implementing one action is incompatible with implementing any other. For example, in the family car problem, purchasing one automobile model precludes the simultaneous purchase of another. This mutual exclusion of actions stems from their being conceived in a manner that captures the decision in a comprehensive fashion – e.g., the location of the entire highway or the complete distribution of subsidies.

On the other hand, when the action is identified by, for example, a loan application, a "sector-application" couple (Ex. 7), or an individual requesting a subsidy (Section 5.1.1), several actions in the model could be combined for joint execution in the setting of the comprehensive decision. This fragmented character does not imply that all the actions of the model are compatible for a given phase of study. Eventually, this will have to be the case, but only for certain actions and in a later phase of the study.

DEFINITION 5.1.2: An ***action is comprehensive*** *if it is exclusive of every other action introduced in the model when executed;*[1] *otherwise, it is* ***fragmented****.*

Example 3: Agricultural Development (from Section 3.2)

Suppose that an agricultural plan can be characterized by a certain number of variables denoted x_1, ..., x_m, where x_h can represent:

- the area assigned to a certain type of production (citrus fruits, rice, vegetables, ...): Given that it is possible to grow several crops on the same parcel of land during a year and that the yield would depend on the order in which the crops are planted, one must also specify the first crop planted and, for the other crops, the crop that was planted immediately before it. It might also be necessary to specify the type of land

[1] *Execution is considered from the viewpoint of its contribution to the comprehensive decision.*

on which the crop is planted. Thus, x_1 might denote an area of dry land planted with vegetables for the first harvest, while x_2 might denote an irrigated area of land planted with vegetables that had previously been planted with rice;
- the number of dairy cows raised on the farm or in pastures;
- the number of cattle, pigs, and other animals, raised for meat in stables or fields.

Under certain conditions, the vector $a = (x_1, ..., x_m)$ defined in this way would completely specify an action that we shall call a plan. For such a plan to be the application point in the decision aiding study, we shall assume that within the PDS considered:

- the demand is fixed;
- the yields are known as functions of the crop conditions;
- a product that is imported cannot be simultaneously exported.

Choosing values of the different variables x_h for a given product, therefore, determines the production, which in turn determines imports and exports. Under these conditions an action a would be identified by a production plan, which is a vector $\underline{x}$[2] of the individual x_h's – that is,

$$a = (x_1, ..., x_h, ..., x_m) = \underline{x}.$$

In this case, we are dealing with a comprehensive conception: Each action (i.e., each plan, whether or not it is realistic) is comprehensive by definition (see Def. 5.1.2). To define a plan that could be implemented, the components of $\underline{x}$ would have to satisfy various physical conditions (e.g., space limitations, demographic development, common agricultural practices employed). For the rest of this example, we shall assume that all the conditions can be written as linear inequalities (for more details, see Fayette, 1975) and shall denote the set of solutions to the linear system by A.

(continued in Section 6.1.1)

Example 6: Research Projects (from Section 3.4)

In this case, the object of the decision aid is to help a committee confronted with a repetitive decision. We shall assume that the recommendation is to lead to a methodology that would allow the committee to decide as objectively and efficiently as possible, given the objectives and priorities defined by the upper management of *O*. Even though *O*'s research and development guidelines have only been stated in general terms, it is clear that the application point of the decision aid is the research task. This is the level at which the funding decision is made, and the methodology must, therefore, operate at

[2] *We shall use underlined letters to denote vectors.*

this level. A different system might lead to considering the methodology at a different level (for more details, see Le Boulanger and Roy, 1970).

Each task submitted to the committee, whether it concerns a continuing project or a new one, constitutes an action, since it satisfies the autonomy requirement of Section 5.1.1. In this case, however, an action cannot be considered independently of other actions: A task might duplicate, be incompatible with, or, on the contrary, complement some other action. It is, therefore, a fragmented conception.

Let us assume that the committee has required the divisions of *O* to review all actions which they submit for funding consideration, and that a specific format for submission (e.g., requiring the divisions to answer precise questions) has been developed and accepted by the organization. We shall let A denote the set of actions that have passed this initial review and that must be considered by the committee whenever it meets.

(continued in Section 6.1.3)

These examples bring up a somewhat subtle aspect of modeling: What level of detail is required for an action to be operational? When the formal representation of the action is a vector (see Ex. 3), for example, the analyst must choose among several possibilities, such as the number of components in the vector and the specific nature of each. Similarly, when the action is defined as an application (see Ex. 6), he must precisely describe the elements that the application would include. The choices made should depend on the degree of detail needed to compare actions. Each action must be identified in such a way that all the features that distinguish it from the other actions in the context of the problem are formally recognized.

The examples of this chapter indicate that one of the following two conceptions will often emerge naturally after the PDS is defined:

- A **comprehensive conception** is one in which each action is comprehensive. The term "alternative" is often used for an action in this case.
- A **fragmented conception** is one in which each action is fragmented. The action is only part of an incompletely specified alternative.

Appropriately combining fragmented actions can lead to mutually exclusive configurations that form comprehensive actions. The fact that these new actions are comprehensive does not necessarily make them operational application points in the decision aid, however. The ability to use them meaningfully in the decision process may be compromised in the transformation. Moreover, these configurations are more complex than simple juxtapositions of the original fragments. Finally, the number of comprehensive actions that would result when combining the fragmented actions in all possible ways increases the number of actions considerably and complicates the analysis of feasibility (see Section 5.2). The analyst must consider these aspects carefully before selecting a comprehensive conception that does not arise naturally.

Finally, let us consider again the influence of the PDS on the specification of the action (Def. 5.1.1). Specifically, let us consider how the PDS can affect the conception of the critical point to which the current phase of the study will lead. In the subsidy example, we said that each beneficiary could form the basis of an action by defining the amount of subsidy each would receive relative to some norm. The decisions that help place the different beneficiaries, in a qualitative way, somewhere relative to this norm lead to a new PDS for which the action can be represented as a vector describing the comprehensive subsidy distribution among the different beneficiaries.

Such a comprehensive conception follows from a conception that was initially a fragmented one. The Agricultural Development example (Ex. 3) illustrates the possibility of the reverse: After using the comprehensive conception described above to determine the general directions to follow, the analyst can turn to a fragmented conception to guide the determination of the most important investments.

One could also imagine a **mixed conception** that combines the comprehensive and fragmented conceptions in the same phase of the analysis. Eventually, however, the conceptions would probably be considered in a hierarchical fashion, as was done above. We, therefore, only mention this mixed conception for the sake of completeness, since we have not come across any practical example that justifies discussing it further.

5.2 THE SET OF POTENTIAL ACTIONS

5.2.1 Delimiting the set of possible actions

DEFINITION 5.2.1: A ***potential action*** *(or candidate) is an actual or dummy action temporarily judged as being realistic by at least one actor or assumed as being such by the analyst. The set of potential actions on which the decision aid is based during a given phase of the study will be denoted A.*

The concept of feasible (achievable or possible) solutions traditionally used in operations research pertains to a stronger definition that excludes any temporary character. Therefore, we prefer to use the qualifier "potential" in our definition to emphasize this aspect. Separating what is assumed to be realistic from what should be treated as unrealistic within the PDS – that is, determining the potential actions to be temporarily considered, even if they are to be excluded later – can pose somewhat tricky boundary problems for the analyst. We have already caught a glimpse of these difficulties: Which sector-application pairs in Example 7 (Section 5.1.1) should belong to A? In Example 3 (Section 5.1.2), what values should be given to the coefficients of the linear constraints defining the set A of plans assumed capable of being implemented? In large part, these problems stem from the fact that defining the membership conditions of A is inevitably somewhat arbitrary. But, even when this difficulty is overcome, A can only be **finally determined** when the two stability conditions presented below are satisfied (as would be the case in Ex. 7 of Section 5.1.1).

CONDITION 5.2.1 (Internal Stability): Given its internal conception, the study phase is such that there should be no reason to reconsider the initial definition of A, except perhaps in a minor or marginal fashion. The decision aiding methodology can consider the set A as fixed and not be concerned with generating possible revisions of A.

When this internal stability condition is not satisfied, a distinction can be made in the same phase of the study between the initial set A_0 and the sets A_1, A_2, ..., that can be derived from it by successive revisions. The two most important causes of revision are the following:

- The boundaries that characterize the potential actions are recognized as being arbitrary, justifying revisions as the study phase progresses. Systematic adjustments based on results obtained along the way might even be a dominant feature of the decision aiding methodology.
- The results obtained with the initial set A appear insufficient and motivate investigation of new potential actions. In this case, it is no longer an issue of a simple rectification of the boundary. It becomes necessary to rethink the very basis used to establish A. Both types of revision can occur in Example 3.

In this way, the study often feeds back into its original conception by causing an evolution in some of the initial data and assumptions. Condition 5.2.1 above concerns one aspect of this feedback that the analyst must consider at the beginning of each phase of the study.

CONDITION 5.2.2 (External Stability): The recommendation expected from the study phase is based on a set A that, because of the external context, has a certain permanence; it is subject only to minor or marginal changes. The decision aiding methodology can consider a set A that can possess a durable, exhaustive definition and does not have to consider a possible transitory character of A.

The external stability condition would not be satisfied if the recommendation must be adapted to a set A that is expected to exhibit important changes due to the natural effects of its environment (e.g., creation and disappearance of potential actions in Example 6).

DEFINITION 5.2.2: The set A is called:
- ***stable*** *if the conditions of internal (Cond. 5.2.1) and external (Cond. 5.2.2) stability are assumed to be fulfilled;*
- ***evolving*** *if it is not stable;*
- ***fixed*** *if the condition of internal stability (Cond. 5.2.1) is assumed to be fulfilled;*
- ***revisable*** *if it is not fixed;*
- ***permanent*** *if the condition of external stability (Cond. 5.2.2) is fulfilled;*
- ***transitory*** *if it is not permanent.*

Figure 5.2.1: Possible cases relating to the of stability of A

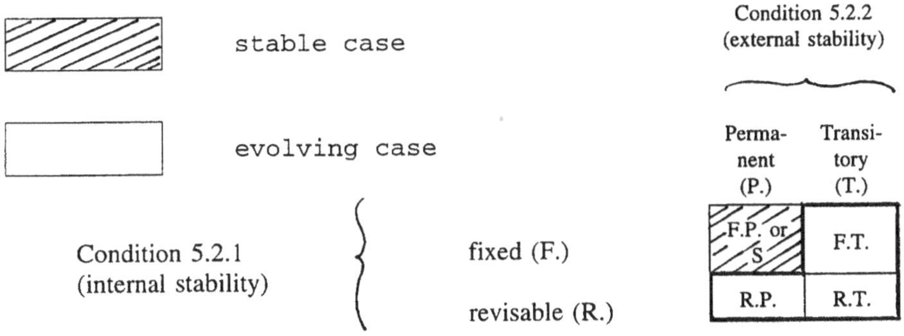

Figure 5.2.1 summarizes these different cases, each of which could obtain in either a comprehensive or fragmented conception (Section 5.1.2). It illustrates that A can evolve in three distinct ways.

Before introducing the final examples of this chapter, it is important to warn the modeler against a tendency to force a comprehensive and stable conception on the set A. Doing so may overly constrain the decision aiding effort. This abuse of stability, as well as that of the comprehensive conception, are generally manifestations of an instrumental bias (Section 2.2.6).

5.2.2 Examples

We now continue with the reference examples to illustrate the above definitions. The reader can refer to Table 6.2 to see which cases correspond to the examples.

Example 2: Commuter Rail Line (from Section 3.1)

An action is defined by specifying the various design elements (stations, capacity, ...) as well as the choices available for each of the elements. A consistent combination of these choices would specify what can be called a (possible) variant for this new rail service. After considering these elements, the analyst comes to the following conclusions (see Fig. 5.2.2):

a) There are two possibilities for the terminal in suburb $\mathcal{B}$:

- end the line at B_0 and build a terminal;
- continue the line to B_1, which would require building a new rail segment but would simplify the development of the terminal;

b) There are several ways to develop the line, depending on whether the line is to:

- be operated up to the central station E_0 (by reclaiming existing infrastructure);
- be operated (on existing infrastructure) only to E'_0, where the extension from E'_0 to E_0 is left as a future possibility (E'_0 is less centrally located and much more poorly connected to the public transportation system serving $\mathcal{E}$);
- be operated to a station E_1 on the outskirts of $\mathcal{E}$, requiring modifications at C_1;
- be connected at C_1 to the rail system serving $\mathcal{E}$ by the intermediate E_1, but with those passengers travelling from B_0 or B_1 to E_1 being required to change at C_1;
- be connected at C_2 to the public transportation system serving $\mathcal{E}$ via E_0, with passengers again being required to transfer.

The analyst decides to consider only these five options during this first phase of the study. However, each option can be considered representative of a family of solutions that differ in secondary design considerations.

c) The two solutions that end at E_0 and E'_0 use the right-of-way of a currently operated line L. There are two possibilities here:

- add tracks and continue present operations on L;
- do not add tracks and discontinue present operations on L.

d) The set of possible stations is the same along segments common to all the solutions, and the analyst has developed a temporary list of stations assumed to be operating in the first stage of implementation. As for line segments that differ among solutions, only stations I_1, I_2, and I_3 (which are on the segment connecting E'_0 to E_0) need to be considered explicitly; the analyst considers only two possibilities concerning these three stations (see Fig. 5.2.3).

e) Given the objectives presented in Section 3.1, certain assumptions must be made on the actual train operations if the variants are to be compared. The analyst believes that he should make at least two hypotheses in the beginning:

- H_1: All trains stop at all operating stations; there are eight trains per hour during peak periods and only four trains per hour during off-peak periods;
- H_2: There are six trains per hour that stop at all operating stations during all hours of service; in peak periods there are an additional six through trains per hour that leave $\mathcal{E}$ for a station J (which is the same in all the options) and then proceed directly from J to B_0.

The analyst, therefore, specifies 36 variants (see Fig. 5.2.3) for this first part of the study. In a comprehensive conception, each specifies an action. Given the limited amount of time and resources available, the analyst decides to reduce the initial set A_0 of potential actions from the 36 variants to 16 sufficiently different ones. These 16 variants are underlined in Figure 5.2.3. He believes that the results of his initial investigation of A_0 will allow him to eliminate some of these 16 variants. This investiga-

Figure 5.2.2: Example 2 transportation network and possible infrastructure

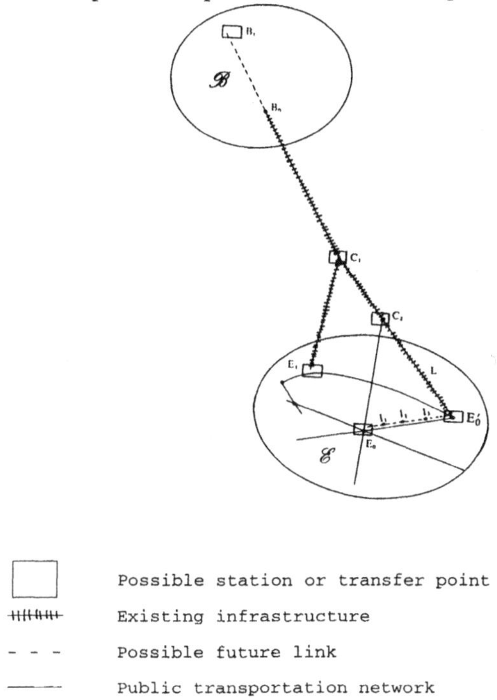

tion should also allow him to determine which of the 20 variants of Figure 5.2.3 not in A_0 warrant closer investigation. We shall call A_1 the set of actions that remain after adding these variants to those in A_0 that were not eliminated in this initial investigation.

To improve the quality of his study and his eventual recommendation, the analyst imagines other revisions that would allow him to refine and to diversify the options described in b), d), and e) above.

Condition 5.2.1 would not hold when formulating the problem in this way. On the other hand, Condition 5.2.2 would hold, since the external context appears stable and correctly analyzed in the near term.

(continued in Section 6.1.3)

Example 11: Product Composition (from Section 3.7)

In this example, we can write:

Figure 5.2.3: Arborescent organization of a first-cut set of variants in Example 2 (underlined variants indicate initial potential actions)

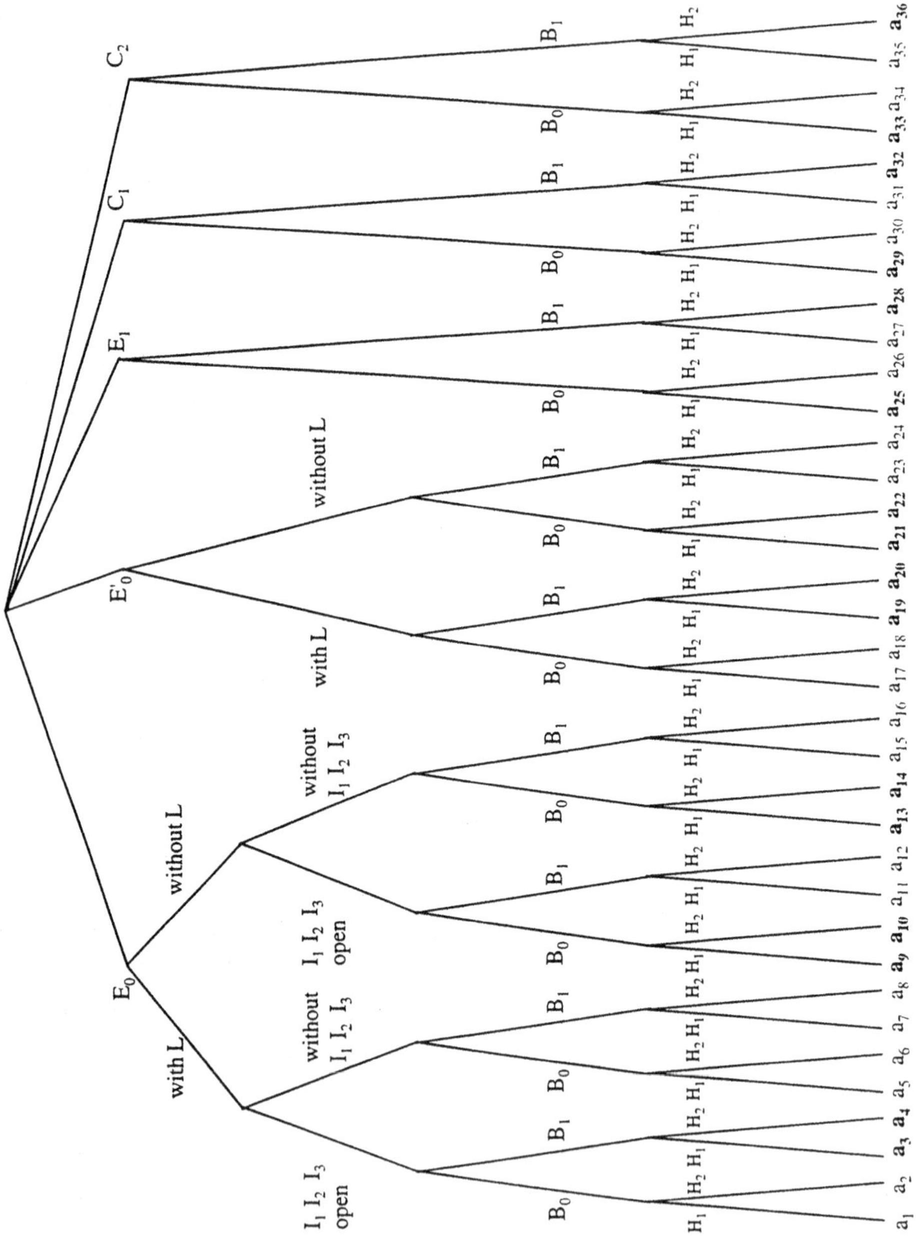

$$a = (x_1, ..., x_h, ..., x_m) = \underline{x}^3,$$

where x_1, x_2, ... represents the amount of stearic acid, sulfur, ..., in a product composition (on the basis of 100 sheets of rubber). We shall denote by:

$$y_k = p_k(\underline{x}),\ k = 1, ..., 12,$$

the performance level obtained by a mixture $\underline{x}$ on property k (external heat resistance, fracture stress, ...), where there are 12 properties considered. The analytical forms of the p_k functions are known, but they only hold over a limited region in $\mathbb{R}^m$ space describing combinations of $\underline{x}$.

To define manufacturer $\mathcal{F}$'s production costs for a unit of product with composition $\underline{x}$, it is convenient to introduce an additional performance function p_0:

$$y_0 = p_0(\underline{x}).$$

In this comprehensive conception, the set A of potential actions (i.e., realistic compositions $\underline{x}$) corresponds to a region of $\mathbb{R}^m$ space that is determined for the most part by conditions of the form (see Fig. 5.2.4):

$$x_h^l \leq x_h \leq x_h^u, \text{ for } h = 1, ..., m$$
$$y_k^l \leq y_k \leq y_k^u, \text{ for } k = 0, 1, ..., 12, \qquad \text{(r 5.2.1)}^4$$

where l and u, respectively, denote given lower and upper bounds. Other mathematical conditions could also be added to represent technological and economic relations.

It should be clear that the analyst will not be able to consider the set A as being permanent. Because of the varying characteristics requested by the client, different components and properties may or may not be considered and cost may or may not be considered a constraint. This variability gives A a transitory character.

As for internal stability, it is useful to look at two cases:

1st case: The relationship between manufacturer $\mathcal{F}$ and the client is such that the set A can be considered fixed once the client's request has been received and modeled as in (r 5.2.1).

2nd case: The relationship between manufacturer $\mathcal{F}$ and the client is such that the set A_0, defined in (r 5.2.1) after receiving the client's initial request, can be revised to reflect changes in the client's wishes that would, themselves, result from suggestions made by

---

[3] *Recall that we are using underlined letters to denote vectors.*

[4] *We shall use r followed by a number indicating the section and its sequence in the section to denote a relation that will be referred to later.*

representatives of $\mathcal{F}$. Specifically, initial results obtained when using A_0 may indicate interesting compromises between minimizing the costs and maximizing the performance levels of various properties.

Figure 5.2.4: Representation of A in Example 11 when considering three products

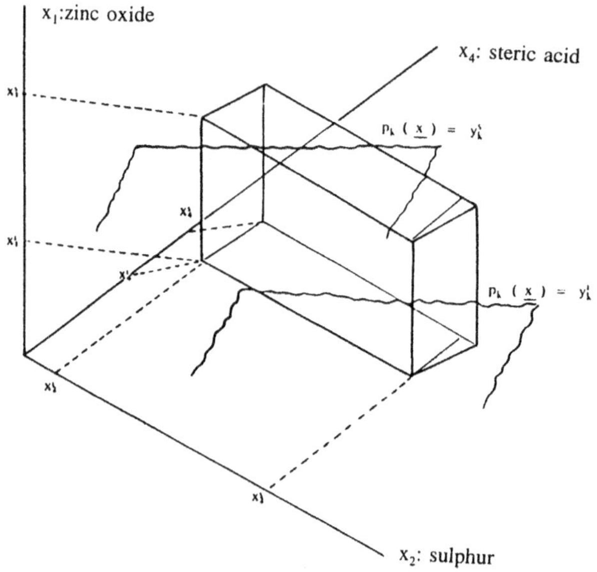

(continued in Section 6.1.1)

Example 9: Engine Assignment (from Section 3.5)

At time t, let H_t be the set of tasks for which execution has not begun and for which the characteristics are known. Among these tasks, it is useful to distinguish:

- those to which the dispatcher has already assigned an engine and for which the assignment is not being reconsidered (owing, for example, to the state of the assigned engine or changes in the task characteristics); we shall use J_t to denote the subset of H_t containing these tasks (most of the tasks for which $t_i < t + 60$ belong to J_t);
- those to which the dispatcher has not yet assigned an engine or for which the assignment is being reconsidered; we shall denote this subset I_t ($I_t = H_t \backslash J_t$[5]).

Now let P_t be the set of tasks such that the engines that performed them have not yet begun their next tasks. Of these tasks, those for which engines have already been reassigned to tasks in J_t form a subset R_t. Let us call $Q_t = P_t \backslash R_t$.

The decision aid might revolve about a conception in which the tasks of I_t and the engines that are free after performing tasks in Q_t are considered as the application point. Nevertheless, one must not lose sight

[5] "\" *indicates the set difference: In this case, the elements of I_t are those in H_t that are not in J_t.*

of the fact that the reassignment of engines which are free after performing tasks in R_t to tasks in J_t can still be considered. Without entering into details on modeling A, simply note that a task in I_t can be the departure point of a somewhat more complex action on the part of the dispatcher. Suppose that the analyst has his reasons for handling this complexity by forming a hierarchy consisting of the three following levels, where the dispatcher will only go to level n if level n – 1 will not lead to a satisfactory decision:

Level 1: Try to assign to task i an engine k freed from a task $q \in Q_t$ at time $t + \varepsilon$ ($\varepsilon = 0$ if q is finished at time t) while satisfying the constraint:

$$t + \varepsilon + D_k(E_q^*, E_i) < t_i. \qquad \text{(r 5.2.2)}$$

Level 2: Telephone the superior to determine allowable deviations from the time t_i and the state E_i; then, try to assign an engine freed from a task of Q_t that satisfies a relaxation of constraint (r 5.2.2) according to the deviations received.

Level 3: Try to find an assignment by reconsidering one or several of the assignments already decided upon and, if necessary, by consulting the superior on allowable deviations from times t_j and states E_j of tasks $j \in J_t$ that can be revised for later assignments.

Note that a decision made at Level 2 relative to task i could consist of making:

- an assignment that (*a posteriori*) could have equally well been made at Level 1, in that the assigned engine k satisfies t_i and E_i; consulting the superior could be explained by the fact that the dispatcher believes it prudent to reserve k for another task if another solution could be found for i;
- no assignment (without going to Level 3) at time t; this situation can, nevertheless, be accompanied by a modification of t_i proposed by the dispatcher and accepted by the superior.

Under these conditions the analyst might seek to provide the dispatcher with a means of using the computer to choose the task(s) that should be given highest priority at time t. In addition to indicating such tasks, the model might also provide a list of the engines to be considered for a given task in a decision of Level 1, 2, or 3.

At any given moment the dispatcher can only concentrate on one or at most a few of the tasks in H_t in coming to a decision at Level 1, 2, or 3. Moreover, there are always new tasks being requested and completed. For these reasons, the set A used in the model results from a fragmented conception; it would also appear transitory. The possible violation of the condition of external stability is all the more highlighted by the fact that the characteristics of t_i and E_i can be modified exogenously.

The condition of internal stability does not appear to be satisfied either: The dispatcher will reconsider the characteristics of a task (therefore, of the definition of A) when she finds herself at Level 2 or 3; that is, having analyzed the assignment problem implied by this task, she might find it necessary or desirable to modify the implementation of the assignment. The model must be able to address such potential modifications if it is to become operational.

(end of Example 9)[6]

[6] *translator's note: In the original French version of this book, Example 9 is continued in Sections 6.1.2, 8.2.4, 9.2.2.1, 9.3.4, and 11.4.1.*

Chapter 6

PROBLEMATICS[1] AS GUIDES IN DECISION AIDING

SUMMARY

In Section 6.1, we define and illustrate four reference problematics – P.α, P.β, P.γ, P.δ. The objective of P.α is to aid the decision maker by the choice of a subset that is as small as possible so that a single action can eventually be chosen. This subset contains "best" actions (optima) or, perhaps, "satisfactory" actions (satisficing solutions). The result of P.α is a choice or a selection procedure.

The objective of P.β is to aid the decision maker by a sorting that leads to an assignment of each action to a category, where the categories are defined beforehand as a function of certain norms that deal with the ultimate fate of the actions that will be assigned to them. The result of P.β is a sorting or an assignment procedure.

The objective of P.γ is to aid the decision maker through a ranking that is obtained by placing all, or simply the "most attractive," actions into equivalence classes that are completely or partially ordered according to preferences. P.γ results in a ranking or an ordering procedure.

The objective of P.δ is to aid the decision maker by developing a description of the actions and their consequences in appropriate terms. It results in a description or a cognitive procedure.

In Section 6.2 we discuss how the problematic chosen for a given phase of the analysis can correspond to one of the four reference problematics, to a sequence of two or more of the problematics, or to a mixed problematic. We illustrate the last two cases with new examples.

6.1 THE FOUR REFERENCE PROBLEMATICS

Given the set A of potential actions, the analyst must now determine in what terms he will pose the problem. What types of results does he envision, and how does he see himself fitting into the decision process to aid in arriving at these results? Towards what will he direct his investigation? What form does he foresee his recommendation taking? Should he envision his recommendation as a tactical one, analyzing only the actions of A and their consequences, or should it be more strategic, attempting to develop a methodology that could be used repeatedly, or even automated? We use the word problematic to describe the analyst's conception of the way he envisions the aid he will supply in the problem at hand based on answers to these questions.

---

[1] *translator's note: We considered translating this term as problem statements, problem types, or problem formulations, but felt that these could give the wrong impression. After discussion with several researchers in this field, we decided to remain close to the the original French word "problématique," even if it seems like jargon and sounds somewhat awkward, but which we feel will avoid misunderstanding.*

To help answer these questions we propose four general problematics to serve as references. Section 6.1 is devoted to defining and illustrating these problematics. Their use is closely associated with the decision process development state (PDS; Def. 4.1.2) and with the idea of the next critical point to which the PDS will lead. It is possible that none of these problematics fits an actual problem stage exactly; the analyst may need to refer to more than one of them. We discuss such issues in Section 6.2.

6.1.1 Choice problematic P.α: Help choose a "best" action or develop a selection procedure

This most traditional of problematics presents the problem in terms of a "best choice" and forms the basis for traditional optimization procedures (see Roy, 1976). According to the definition that we propose below, however, the optimization problem is only a special case of this choice problematic.

DEFINITION 6.1.1 The ***choice problematic P.α*** *presents the problem in terms of choosing one "best" action – that is, of directing the investigation towards finding a subset A′ of A, as small as possible, that will enlighten the decision maker as to what should be the outcome of the next critical point of the analysis, while keeping in mind that A might evolve. This problematic results in a recommendation or simple participation that either: indicates a decision that should be taken; or proposes a perhaps automated selection-based methodology that can be used repeatedly to identify best actions.*

In this problematic (see Fig. 6.1.1) the elements of A are compared to each other so as to eliminate as many actions as possible. The ideal would be to eliminate all but one action that would be at least as good as all those which had been eliminated. Obtaining one such action (an optimum) may be impossible, however, or require rather arbitrary assumptions. For example, the set of potential actions A may evolve or be revised with time. Or, it may not be possible to measure the consequences of actions with enough precision to settle on a single action. Moreover, the multiple value systems impinging on the problem may make it impossible to determine a "best" action.

Figure 6.1.1: Representation of a choice problematic outcome

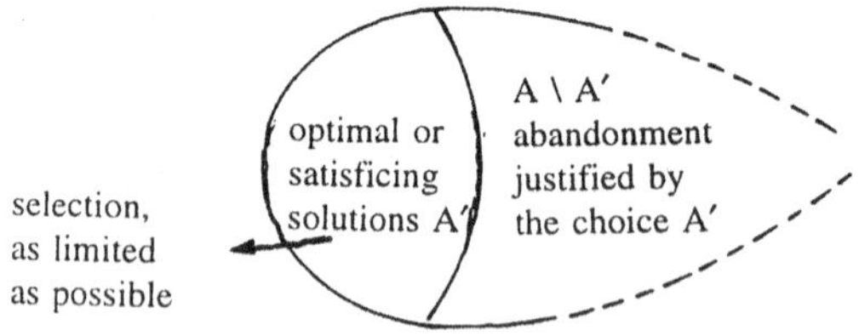

In any case, the choice represented in problematic P.α concerns a subset A′ of A. A′ should be as small as possible and be such that for any action in A\A′[2] the decision maker would prefer some action in A′, or at least such that the actions in A′ are considered good enough to eliminate those of A\A′ from further consideration.

When A′ contains more than one action, its elements might all be considered equivalent and better than any action not in A′. That is, they could be considered multiple optima. There are times, however, when the elements should not be considered equivalent. For example, each element may represent a unique optimum based on a different value system. Or, the elements may be the product of explicit attempts to form compromises among very different technical options and, therefore, be difficult to compare. Similarly, the imprecision, uncertainty, or inaccurate determination in predicting their anticipated consequences (Section 8.2) may be great enough that one would be unwilling to declare the elements of A′ as multiple optima. Following Simon (1981), we shall use the term "satisficing" action to denote any element of A′ that cannot be considered an optimum.

Example 1: Site Selection (from Section 3.1)

Assume that the first phase of the study resulted in: i) a survey of all the available sites in the region considered acceptable for the institute, and a summary of the most important advantages and disadvantages of each site; ii) a preliminary analysis of the sites, eliminating those that would not be considered worthy of further analysis. Assume also that this preliminary analysis was a coarse one: Discussions with the various stakeholders about the possible anticipated consequences of the sites were based on limited and rather sketchy data, but still led to rejecting all but 5 of the initial 17 candidates as incompatible with the institute's strategic development plans (see Fig. 6.1.2). Let the set of remaining actions for the second stage of the analysis be $A = \{a_1, a_2, a_3, a_4, a_5\}$, where a_i corresponds to building the institute at site i.

For this phase of the analysis, the problem is presented as "choosing a best site." More specifically, the analyst is to determine whether one of these sites would appear significantly better than the other four from the viewpoint of the decision maker, i.e., the director of the institute. If such a site can be found, the analyst's recommendation would take the form of summarizing and synthesizing the elements that the director could use in discussions so as to improve the likelihood that this site would eventually be selected.

The limited resources (technical assistance, time, ...) available to the analyst permit only a preliminary study of the consequences of the five actions. In addition, he knows that the value systems underlying the preferences of the various actors are quite different and that for many of them (including the director) the relative importance that should be given to the diverse consequences is open to debate.

[2] *Recall that A\A′ is the set of all the elements of A that are not in A′.*

Figure 6.1.2: Site locations of the second phase potential actions in Example 1

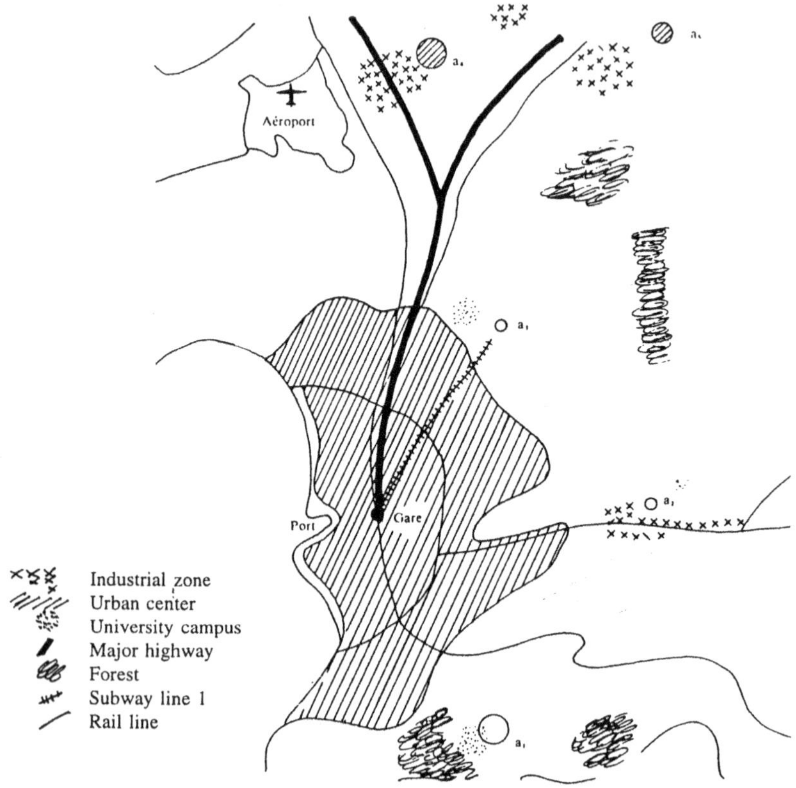

For these reasons, the analyst believes that this choice problematic will only lead to proposing a single site if the reasons for claiming it as optimal are both apparent to the director and difficult for the other actors to attack. Having to choose two or even three sites – which could be equivalent or, more probably, incomparable – seems quite possible. Therefore, in addition to justifying his choice of A′ from A, he must be able to clarify the specific characteristics of each site chosen and the arguments that would allow the director to limit the discussion to only the actions in A′.

(continued in Section 11.3.2)

Example 3: Agricultural Development (from Section 5.1.2)

Problematic P.α also applies to this example. Here, it is a question of determining the characteristics of the recommended plan as precisely and rigorously as possible. But, again, it might be difficult or artificial to define one optimal plan. First of all, the heterogeneous nature of the three objectives pursued by the government of $\mathcal{P}$ – balance of foreign trade, independence of food supply, maintenance of rural employment – make synthesis into a single criterion difficult. (We shall return to this point later.) Secondly, the numerical values given to many of the coefficients of the linear system defining A (a set defining a polyhedron in $\mathbb{R}^m$) may be considered somewhat arbitrary. To search for an optimum in A will almost inevitably lead to a plan that is on the boundary of the polyhedron, especially if the criterion is linear. The plan will, therefore, depend strongly on the arbitrary data that define the boundary and separate it from infeasible plans. The analyst will have to consider revising A if he is to discover other plans that would result from different numerical approximations. He might consider using techniques such as sensitivity analysis or simulation. More than likely, the outputs of these revised investigations would indicate the characteristics of the plan to be recommended. These characteristics would be defined not by a unique vector $\underline{x}^*$ in A, but by a region A′ ($\subset$ A) denoting those plans that can be considered satisficing actions. From the "shape" of this region, the analyst might find somewhat robust elements justifying an interval of values that define characteristics of the plan or a relationship between the values and some other characteristics.

(continued in Section 8.1.5)

Example 11: Product Composition (from Section 5.2.2)

In both the first and second cases, the upper management of $\mathcal{F}$ expects the analyst to propose a methodology based on a selection procedure for "best mixtures" that could be used repetitively. This is, then, another example of problematic P.α, but this time from the strategic perspective discussed in the introduction to this section.

In the first case, $\mathcal{F}$'s upper management (the client of the decision aiding effort) might wish to automate to a large extent. This poses no major difficulties *a priori*, since the methodology would reside within the technical division, which is the decision maker as defined in Section 2.2.2. The problem is well defined, and the information is available.

In the second case, the issue is one of choosing elements based on a set of initial actions A_0 that will allow $\mathcal{F}$'s marketing representatives to propose modifications or more details related to the initial request. This could generate other interactions between the manufacturer and the recipient of the product that could, in turn, lead to subsequent revisions of A_0. These interactions should help the technical division make the soundest possible decision corresponding to the wishes of the company on the eventual product composition.

(continued in Section 9.1.1)

6.1.2 Sorting problematic P.ß: Help sort actions according to norms or build an assignment procedure

This problematic arises when a set of potential actions must be sorted according to the results of a test: a credit application test, certification test, diagnostic test for a patient, ... (see, for instance, Moscarola and Roy, 1977; Roy, 1981; Massaglia and Ostanello, 1991). We encountered this problematic, without calling it such, in the first phase of Example 1 (Section 6.1.1). We now define and illustrate it with examples.

DEFINITION 6.1.2: The ***sorting problematic P.β*** *presents the problem in terms of placing actions in categories that are defined in terms of the eventual fate of the actions – that is, of directing the investigation towards determining an assignment of actions in A to these categories based on norms related to the intrinsic value of the actions, while keeping in mind that A might evolve. This problematic leads to a recommendation or simple participation that either: advocates accepting or rejecting certain actions or furnishing more complex recommendations (depending on the categories); or proposes adopting a perhaps automated assignment-based methodology that can be used repeatedly to place actions in categories.*

In this problematic (see Figure 6.1.3), one tries to establish characteristics such as: "certainly good," "certainly true," "probably satisfactory," "definitely bad," "definitely false." In general, the issue is one of characterizing membership conditions for diverse categories (e.g., actions to be accepted, rejected, or put aside until further information can be obtained) that partition A for use in eventual recommendations. To be useful to the decision maker, each category must be defined as a function of the treatment that will be given to the actions in the category. A category could be developed to accept those actions that do not seem to fit any other category. It is this type of partitioning that forms the basis for scholastic entrance exams or medical tests, such as X-rays or blood tests.

Figure 6.1.3: Representation of a sorting problematic outcome based on predefined categories that segment A

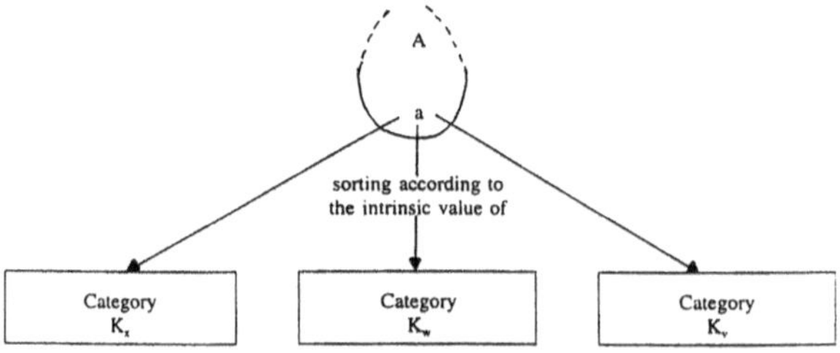

The sorting considered in problematic P.ß is one of assigning each action to exactly one of the categories developed to direct the decision. Each of the basic categories must possess an intrinsic definition, i.e., a definition that does not refer to the other categories. Note that the dichotomy (A′, A\A′) that results from problematic P.α does not represent a partition into categories for selection as defined here, since the dichotomy is based on a comparison of the actions rather than an analysis of their intrinsic values.

Example 10: Application Packages (from Section 3.6)

Here, the issue is one of developing an automatic procedure that will reduce the work of the admissions committee by helping it determine those applications that are good enough for admission or bad enough for rejection. By applications that are "good or bad enough," we mean those for which no personal interview or further information is necessary.

In this case, the application point of the decision aiding procedure is a candidate's application package, and A is, therefore, the set of all application packages. A is fixed and not defined by the admissions committee. It is not permanent, however, in that it could be different each time the committee meets.

The procedure that must be developed is one of sorting the application packages. The administration of the academic unit and the admissions committee accept the analyst's suggestion to sort the packages into three categories:

- $\underline{A}_1$: the subset of packages that are strong enough to support immediate acceptance of the applicants,
- $\underline{A}_2$: the subset of packages that are neither strong enough to support immediate acceptance nor weak enough to support immediate rejection;
- $\underline{A}_3$: the subset of packages that are weak enough to support immediate rejection of the applicants.

This segmentation serves as a guide to the admissions committee, and it will only be useful if it can be determined without the committee's intervention. Developing such an automatic process will, therefore, require specifying the characteristics of "good" and "bad" applicants based on the mission of the academic unit. These characteristics will then be used to assign an application package to $\underline{A}_1$, $\underline{A}_2$, or $\underline{A}_3$.

(continued in Section 7.2.4.1)

Example 7: Industrial Development (from Section 5.1.1)

In this example, the recommendation will not be a methodology, but the specifications themselves. The Board of Directors expects the analyst to develop specific guidelines that would allow it to distinguish from the set of 124 "sector-application" couples those which should be promoted in the short term.

The analyst could, therefore, pose the problem as one of sorting out those sector-application couples to which promotional priority should be given. For each action $a \in A$, he must, therefore, investigate in more detail the technical, economic, and political reasons that would argue in favor of developing a given application in a certain sector. Therefore, he directs his efforts towards developing an assignment procedure that sorts sector-application couples into those: i) not deserving promotional priority; ii) for which no conclusion can be reached based on the current information; and iii) deserving promotional priority.

Note that even though this example fits the description of problematic P.ß, the analyst could also have adopted P.γ (Section 6.1.3) as his problematic. Moreover, using P.γ after P.ß, and in the same phase of the analysis, might improve the recommendation: A ranking of the actions in those categories indicating immediate or potential promotional priority could help the Board of Directors in its deliberations.

(end of Example 7)[3]

6.1.3 Ranking problematic P.γ: Help rank actions in order of decreasing preference or build an ordering procedure

Instead of considering an assignment procedure for sorting the sector-application actions in Example 7, the analyst could have considered a procedure to rank the actions according to decreasing preference. In this case, rather than thinking of a "test," we can think of a "competition," and rather than thinking of dividing A into categories based on intrinsic properties of the actions, we can think of comparing actions so that they can be grouped into classes that can be ordered. Unlike the categories of the previous problematic, the classes would group actions that are considered equivalent, at least for those classes near the top of the order. In the cases where only actions in the first few classes would be considered "satisfactory," it seems pointless to try to define lower classes so that each would contain equivalent elements. We shall not, therefore, consider classes at the bottom of the order for which no further refinement is warranted, and these classes cannot be considered equivalence classes[4] in the sense of preferences.

DEFINITION 6.1.3: The ***ranking problematic P.γ*** *presents the problem in terms of ranking the actions of A (or some of these actions) – that is, of directing the investigation towards determining an order defined on a subset of A so as to be able to determine those actions that could be considered as "sufficiently satisfactory" based on a preference model, while keeping in mind that A might evolve. This problematic leads to a recommendation or simple participation that either: suggests a partial or complete order formed by the classes containing actions considered equivalent; or proposes a perhaps automated methodology based on an ordering procedure (for all or part of A) that can be used repeatedly.*

---

[3] *translator's note: Example 7 was continued in Sections 9.3.4 and 12.5 in the original, French version.*

[4] *For more details, see Section 7.2.2.1 a).*

In adopting this problematic (see Fig. 6.1.4), one tries to use available information as much as possible to compare the elements of A among each other so as to determine classes for the elements in a subset A′ of A and a ranking of these classes. Such a ranking or ordering, which is developed to aid the decision maker, must reflect a certain degree of importance or priority that the decision maker gives to each element of A′. The ranking is designed to help her think about the problem, to guide her discussions with other stakeholders and, more generally, to serve as a framework for approaching the next critical point of the decision process (see, for instance, Roy and Hugonnard, 1982; Roy *et al.*, 1986; Bana e Costa and Rodrigues, 1990; Pirlot, 1990; Roy *et al.*, 1992.

The ordering procedure considered in this problematic consists of assigning a "rank" to each action in a subset A′ ⊂ A, where two actions are assigned the same rank whenever the data do not allow a distinction between them with respect to the next critical point of the analysis. The ranking of the classes allows an ordering, either complete or partial (see Fig. 6.1.4), representative of preferences. Unlike the categories of P.ß, the classes of P.γ are not determined from an *a priori* definition. Rather, the significance of a class is relative, in that it depends on its position in the overall ordering. In discussing P.ß, we mentioned that one might define categories for those actions whose intrinsic value is not easily determined. Similarly, one should not necessarily require that the classes of P.γ form a complete order.[5] The poor quality of the data,[6] the conflicting criteria, and the multiple value systems involved might make it difficult or artificial to develop a complete order. In certain cases, a partial order, which does not necessarily position a class relative to every other class, can be quite useful.

Figure 6.1.4: Representation of a ranking problematic outcome

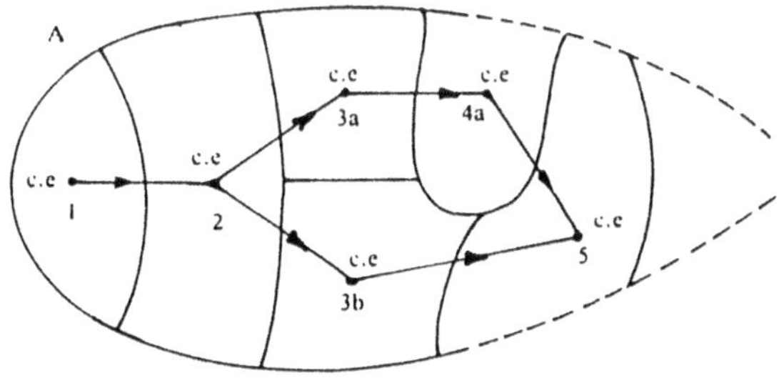

The equivalence classes (EC) are defined relative to each other and ordered according to a not necessarily complete (see Section 7.2.3.2) ranking denoted by the arrows.

Note that the dichotomy (A′, A\A′) obtained in P.α cannot generally be considered a ranking compatible with P.γ. Although one might be tempted to consider A′ as the first

---

[5] *Complete and partial preorders will be rigorously defined later (see Def. 7.2.5).*

[6] *See Section 8.2.*

and only equivalence class, and A\A′ as grouping other equivalence classes into a single category, A′ is not intended to group actions that are equivalent in the sense of P.γ. Moreover, the first class obtained in P.γ is defined neither with the choice of a single best action in mind nor with the objective to be as small as possible (see Definition 6.1.1). P.α and P.γ, therefore, correspond to very different perspectives.

Example 2: Commuter Rail Line (from Section 5.2.2)

The general framework was determined in the first phase of this study. Assume that the original set of possible actions was revised several times (in a manner consistent with P.ß) in collaboration with the client and, in certain cases, with the director's representative. At the end of this first phase, we are left with roughly twenty variants that the decision maker considers worthy of a comparative study based on the five general objectives mentioned in Section 3.1. The study is now ready to move to the second phase.

How the final decision will be made is unclear. Moreover, no one participating in the process is capable of dictating the decision. In addition, there appears to be no single person whose preferences can be imposed over all the others. Even if there were, any individual's preferences would probably be modified as the study progresses and the issues are discussed further. A detailed and rigorous study seeking to pinpoint the best action would, therefore, be inappropriate.

For these reasons, the analyst considers the problem to correspond to P.γ rather than P.α. He feels that it would be most useful to develop an easy-to-use ordering procedure that can quickly place the actions of A into equivalence classes, keeping in mind the questionable quality of the data. Therefore, he considers that two variants should be compared:

i) by using a system of weights that indicate the relative importance of the criteria expressing the five general objectives;
ii) with respect to a specifically stated scenario – i.e., a set of events that the decision maker cannot control, such as the evolution of residential employment and mobility patterns or perhaps the construction of new terminals and transfer points in the next ten years, but that would help describe the possible environment in which the commuter rail line would operate.

The system of weights and the scenario must both be capable of being easily modified so that the decision maker can request analyses of the sensitivity of the ranking to the assumptions made.

(continued in Section 8.2.3.1)

Example 5: Media Planning (from Section 3.3)

The application point of the decision in this example is a periodical, and we shall not reconsider the directives given by the decision maker in agency $\mathcal{A}$. A potential action a is, therefore, completely defined by a specific periodical considered. The issue is whether or not a specific periodical should be included in a combination of periodicals to be studied in Step e).

Let A be the set of periodicals with which the agency works. It is, therefore, the set of potential actions and is a fragmented conception. It can be considered stable in the sense that the agency only rarely modifies this set of periodicals.
The analyst is asked to improve Steps a), b), c), and d) of the methodology described in Section 3.3. After having discussed the matter with the relevant personnel, he decides to develop an ordering procedure that ranks the periodicals in A in order of decreasing importance to the advertising campaign. Such a ranking would facilitate discussion and assist in determining the combinations that would serve as inputs to Step e). It might also form the basis of a systematic procedure that would lead to these combinations. An example of such a procedure might be:

- choose a periodical in the class that is ranked first;
- eliminate from future consideration this periodical and all those that would be redundant with it (in terms of readership);
- choose a second periodical in the highest class that has at least one remaining element;
- eliminate from future consideration this periodical and all those that would be redundant with it;
- select a third periodical;
- ...
- stop when the selection of a new periodical would surpass the budget or be incompatible with other constraints defined *a priori*.

In this example, the ordering procedure would be used repeatedly. Step b) requires that the methodology in which it is embedded employ a preselection to make it more manageable. The number of elements in A could easily surpass one hundred, but most of them could be quickly excluded as uninteresting in a given campaign p. Therefore, it would be a waste of time to rank A completely. Step b) of the methodology has, therefore, been developed to determine a subset A'_p of A that would group all of the periodicals of potential interest to campaign p. Only the elements of A'_p need to be ranked, and according to the director's experience there would be no more than a few dozen of these elements.

(continued in Section 8.1.5)

Example 6: Research Projects (from Section 5.1.2)

The analyst intends to propose a methodology based on an ordering procedure to the committee. This procedure must lead to a ranking of the research tasks in A before each committee meeting, and the ranking must conform to the policies of upper management.

Such a ranking would be useful in defining the order in which the committee considers the research tasks for funding. The order would be important due to difficulties associated with complementary or redundant tasks. The committee might even seek a consensus on such a ranking, although the consensus ranking would probably be different than the ranking initially developed. Such a consensus would help clarify priorities and demonstrate to upper management the impact of the budget on the tasks eventually funded.

This type of methodology seems well suited for decisions made at the end of the year. Would it also be appropriate for those decisions that would have to be made during the year (see Section 3.4)? Consider the case of a research task a_t for which a funding request occurs in the first or second quarter of year t. Let A_{t-1} be the set of actions considered at the end of year $t - 1$ that would be financed from the budget in year t. Applying the ordering procedure on the union of set A_{t-1} and a_t would provide the ranking of a_t if t had been submitted for review several months earlier. If a_t is highly ranked in this set, the committee could decide to finance it immediately if the budget in year t allows. On the other hand, if it is poorly ranked, the committee would immediately reject the request. In the other cases, the committee would do well to delay the decision until the end of year t.

(continued in Section 8.1.5)

6.1.4 Description problematic P.δ: Help describe actions and their consequences in a formalized and systematic manner or develop a cognitive procedure

Often the client can expect nothing more than a somewhat complete and rigorous description of the actions envisioned by one or more decision makers and, perhaps, the consequences of these actions that should be considered. Sometimes a systematic means to obtain such a description, rather than the description itself, would be sufficient. In this case, the analyst provides a more rudimentary type of aid, one that poses rather than solves the problem. In fact, there are many decision problems falling within the framework of this fourth problematic. These problems can arise at the level of the individual, the firm, or the government, as illustrated in the family car, loan application, and highway location examples provided in the introductory part of this book.

DEFINITION 6.1.4: The ***description problematic P.δ*** *presents the problem in terms of describing the actions of A and their consequences – that is, of directing the investigation towards making the information related to potential actions explicit so as to help the decision maker discover, understand, or evaluate the actions, while keeping in mind* *that A might evolve. This problematic leads to a recommendation or simple participation that either: systematically and formally describes the actions and their consequences in qualitative and quantitative terms; or proposes a methodology based on a perhaps automated cognitive procedure that could be used repeatedly.*

Although this problematic is included in each of the three preceding ones, it is useful to distinguish it from the others, since it can often be considered a separate contribution made by the analyst in practice.[7] The following chapters further justify its interest. Even though this problematic requires a great deal of creativity, certain well-established procedures can often be used in some components.[8] Systems theory (see Klir, 1972; von Bertalanffy, 1973; Wymore, 1976; Le Moigne, 1977; Chapman *et al.*, 1992) offers general ideas, a conceptual framework, and ways to proceed that can be very useful to the analyst faced with this problem.[9]

6.2 REMARKS ON CHOOSING THE PROBLEMATIC

6.2.1 Factors influencing the choice of problematic

At a given phase of the study, the chosen problematic can correspond to:

- one of the four reference problematics or to a special case of one of them (e.g., the optimization problematic, which is a special case of P.α);
- a sequence of two (or, in rare cases, more) of the problematics P.α, P.ß, or P.γ, uninterrupted by partial decisions that could affect the PDS (Def. 4.1.2);
- a "mixed" problematic, i.e., one that does not fit either of the two preceding cases.

The first case is the most common. It is illustrated by most of the examples in Section 6.1. An example of the second case, a sequence of problematics, can be found in Section 6.1.2, at the end of the discussion dealing with Example 7. There, P.γ follows P.ß. The same sequence would also be applicable to Example 6. The continuations of Examples 4 and 8 below could correspond to other sequences (P.ß–P.α; P.γ–P.α). The continuation of Example 12 below can be used to illustrate the third case.[10]

---

[7] *We return to this point in Chapter 10.*

[8] *Models dealing with multidimensional data analysis (see Bertier and Bouroche, 1975) are often used with regard to this problematic, but we shall not discuss them in this book.*

[9] *translator's note: Additional discussion is provided on page 89 of the original, French version.*

[10] *translator's note: Three categories of factors than can influence the analyst's choice of problematic are treated on page 91 in the French version of this work at this point.*

6.2.2 Examples

Example 4: Water Resource Planning (from Section 3.2)

The way in which the problem is presented makes it difficult to choose between a comprehensive and a fragmented conception. (The fragmented conception would offer more flexibility and be better suited to reaching a compromise.) For either conception, the analyst could hesitate in choosing among several problematics. Although there are several ways to proceed, the options that follow seem reasonable.

In the rest of this example, we shall assume that:

- the numerous technical studies already conducted are sufficient to identify rivers that can be diverted and sites suitable for dam construction either in the mountains or in the plains, as well as the different options available in each project;
- a plan consists of a combination of a subset of the project options coupled with a timetable for implementation;
- the action can be modeled as the irreversible part of the decision – i.e., the part associated with some plan that will not be changed in the first two years;
- the analyst is able to interview the different actors on one or two occasions to determine their positions on certain aspects of the problem or to obtain their approval of certain partial conclusions.

Let A be a set of actions, where each action represents all the elements forming the irreversible part of at least one plan. For simplicity, we shall call the elements of A short-term plans. We shall assume that the set of these short-term plans A has received the approval of the stakeholders – i.e., the IMIVAR, the Ministry of Agriculture, the Jacobins, and the Girondins. The following two phases of the study are anticipated:

1st phase: In the first phase, the analyst considers each group of actors as individual decision makers (see Section 3.2). The objective is to specify the preference system for each decision maker d and partition A into four categories:

- $\underline{A}_1^d$: actions that decision maker d would definitely choose if it were the sole decision maker;
- $\underline{A}_2^d$: actions that d considers inferior to at least one of the actions in $\underline{A}_1^d$, but which it would not categorically reject;
- $\underline{A}_3^d$: actions that d believes to be sufficiently bad to reject immediately;
- $\underline{A}_4^d$: actions that d considers difficult to evaluate and that it cannot easily place in one of the other three categories.

The first phase would, therefore, lead to four partitions, each corresponding to one of the stakeholders. The union of these partitions would be accepted as the starting point for the second phase.

2nd phase: The objective of the second phase is to develop a selection procedure. This procedure would lead to one or more compromise plans acceptable to the different stakeholders. Except in the case where the nature of A makes such a compromise obvious, the four stakeholders would need to furnish more input in this phase, and this input might even lead to a revision of A.

It may be difficult to identify the point that would indicate the end of the first phase. If so, the two phases would be grouped together, forming a sequence of P.ß followed by P.α.

(continued in Section 8.2.3.1)

Example 8: Airport Operations (from Section 3.5)

Let A be the set of the hundred or so fragmented actions that were developed from the results of the previously conducted interviews, surveys, and preliminary analyses. When combining them to form comprehensive actions, a number of inconsistencies were found among the individual elements, and these combinations were excluded. Moreover, implementing certain elements could influence the desirability of certain others.

Because of these dependencies, the airport authorities' request to form a short list based on A turns out to be incompatible with the available information. After hesitating in choosing among several problematics, the analyst obtains the authorities' agreement to conduct a study consisting of the following two phases:

- *1st phase*: Rank the fragmented actions according to decreasing preference (P.γ): here, the fragmented actions would be considered in terms of their own merit, ignoring any interdependence with other actions.
- Discuss the resulting ranking with the authorities of $\mathcal{L}$ to obtain their agreement on a ranking that would act as a departure point for the second phase of the study.
- *2nd phase*: Select (P.α) a small number of "best" combinations of the fragmented actions, where each combination would represent a coherent plan.

From here on, we shall primarily consider the second phase. Regarding the first phase, however, note that an assignment (P.ß) could have been substituted for the ranking (P.γ).

As for the second phase, assume that five or six fragmented actions would generally constitute a plan, and that there would never be more than ten actions in any given plan. Such a plan could be developed by iteratively calling upon P.α to choose actions from A, where A becomes progressively smaller with each iteration. As such, A would be continually revised in this second phase. This process would, therefore, call upon P.α at two levels:

- during the iterative process, when building combinations (i.e., selecting fragmented actions);

– between iterations of this process, when obtaining a small number of "best" combinations.

(continued in Section 8.2.4)

Example 12: Plant Organization (from Section 3.7)

The characteristics that would define a new plant organization are many and overlapping. To define a typical action is, therefore, not easy. We shall ignore the details associated with these difficulties, however, and limit ourselves to discussing their impacts on the definition of A and the choice of the problematic.

Consider an action a that the analyst has precisely defined.[11] No matter how rigorous and applicable he has made his model, the complexity of the action a will create communication problems as soon as it differs even slightly from organizational models with which the various actors are familiar. Moreover, even when a is well understood, it is still not easy for a given actor to decide whether or not it corresponds to an organization of $\mathcal{M}$ that should be considered possible for the time being. The differing value systems of the various actors make the decision even more difficult. How can the set A be defined when its elements can neither be exhaustively enumerated *a priori* nor characterized by rigorous properties? The analyst might develop some tests indicating the relative feasibility of a given action. His model and his operational approach to the problem would, nevertheless, have to acknowledge that A is "open," *a priori*. A large part of the aid offered in this case would, therefore, be in developing and defining organizational frameworks not initially considered and in overcoming biases of individual actors against various frameworks.

In the beginning, the various actors consider five potential actions as typifying the possibilities. They can be summarized as:

a_1: classical continuous assembly line with a one-minute cycle time;
a_2: continuous assembly line with a cycle time much greater than a minute and with inventory;
a_3: assembly by small (not self-managed) work groups with inventory;
a_4: individual assembly (to finish a large component) with inventory;
a_5: assembly by semi-autonomous groups with inventory.

The analyst now plans to work with the different actors to produce other organizational models. His goal is not only to enrich A by adding new actions, but to work towards obtaining the best compromises possible in light of the different value systems present. He also envisions developing one or more rankings, according to the important value systems present, of "the most promising" compromise plans obtained. He knows that these compromise plans will be developed progressively and that he will need to obtain the reactions of the different actors not only to the five reference actions, but to any other action that he or they will propose. So, the set A is partially defined:

– based on performance constraints, even though the various specifications can evolve as the process develops and the resulting feasible frontier cannot be considered permanent;
– by the fact that it contains an enumerated series of precise actions; this series, which is originally limited to the reference actions a_1, a_2, ..., a_5, will grow as the study progresses, thereby illustrating the revisable nature of A.

(continued in Section 7.2.4.1)

[11] *For example, the action could be defined in a way described by Giordano and Suquet (1976).*

6.2.3 Multiple cases

In concluding, we emphasize that neither the conception of A nor the properties relating to its stability exclude any problematic. As shown in Table 6.2, P.α, P.ß, and P.γ can be appropriate regardless of whether A is fragmented or comprehensive, stable or evolving. There are 12 possible cases, 6 of which are further subdivided according to the way in which A can evolve (R.T., R.P., F.T., see Fig. 5.2.1).

What the 12 reference examples demonstrated for P.α, P.ß, and P.γ holds even more strongly for more complex problematics (see Section 6.2.1). It follows that all possible cases that result when combining the nature of A with the type of problematic will not necessarily fall within the framework formed by the 12 cases that we have presented.

Note that in Table 6.2 we have not attempted to address combining two problematics in a sequence. Rather, each line corresponds to the problematic that was explicitly mentioned as offering most interest under the conditions considered, whether it is used as is or in a sequence.

Table 6.2: Characteristics of the continuations of the reference examples

Number, Title, Study Phase	Nature of A		Problematic	Beginning Sections
	Concep-tion[12]	Stabili-ty[13]		
Ex. 1: Site Selection:				
a) 1st phase	C	S	β	3.1
b) 2nd phase[14]	C	S	α	3.1
Ex. 2: Commuter Rail Line:				
a) 1st phase	C	R.P.	β	3.1
b) 2nd phase[14]	C	S	γ	3.1
Ex. 3: Agricultural Development	C	R.P.	α	3.2
Ex. 4: Water Resource Planning:				
a) 1st phase[14]	C	S	β	3.2
b) 2nd phase	C	R.P.	α	3.2
Ex. 5: Media Planning	F	S	γ	3.3
Ex. 6: Research Projects	F	S	γ	3.4
Ex. 7: Industrial Development	F	S	β (or γ)	3.4
Ex. 8: Airport Operations:				
a) 1st phase[3]	F	S	γ (or β)	3.5
b) 2nd phase	F	R.P.	α	3.5
Ex. 9: Engine Assignment	F	R.T.	β (or α or γ)	3.5
Ex. 10: Application Packages	F	F.T.	β	3.6
Ex. 11: Product Composition:				
a) 1st case	C	F.T.	α	3.7
b) 2nd case	C	R.T.	α	3.7
Ex. 12: Plant Organization	C	R.T.	γ (or mixed)	3.7

[12] *F: fragmented, C: comprehensive.*

[13] *S: stable, R: revisable, F: fixed, T: transitory, P: permanent (see Fig. 5.2.1).*

[14] *Unless otherwise indicated, this is the phase considered in the rest of the example.*

LEVEL II

HOW TO DETERMINE PREFERENCES AND ON WHAT BASES

As stated in Chapter 4, the concepts addressed in this part (Level II) are different than those addressed in the previous part (Level I). The investigative effort and the attempts to develop abstractions no longer revolve about the object of the decision but about the consequences of the decisions and the comparisons of these consequences. Regardless of the problematic used, one must address how and on what bases the various potential actions can be compared among each other or compared to reference actions that serve as norms.

When considering such comparisons, the viewpoints of the various actors might conflict as a result of different value systems or different informational systems. This difficulty should not be overlooked, and we try to emphasize as often as possible that the preferences in question are those of an identified actor with an interest in the decision process. We denote this actor by Z.

Actor Z can be an individual, an entity, or a community (see Chapter 1) who takes on the role of the decision maker (see Section 2.2.2). Actor Z's preferences may not be completely formulated, may exhibit internal conflicts, and may not be stable. These characteristics may result from a lack of information, different interpretations of a value system, or divergent value systems. Whatever the reason, the model must be able to tolerate ambiguity, contradiction, and a learning process.

Since it fulfills these requirements, the methodology proposed in this part allows the analyst to construct a preference model that is acceptable to the various actors considered. The traits that are unique to the personality of the decision maker and to her voluntarist positions will eventually be addressed explicitly at Level III.

Part II consists of three chapters treating, in order:

- the "language of preferences": What conceptual bases are required to describe and formalize preferences? What structures appear when modeling these preferences? What are the functional representations that allow them to be manipulated?
- the "foundations of preferences": How can one capture the underlying basis of preferences? How can one address the difficulties associated with the complex, arbitrary, and vague elements involved so that preferences can be developed, justified, and transformed?
- the development of criteria: How can one operationally synthesize complex and imprecise information in the form of criteria? How can one assess the degree to which these criteria can be used convincingly in decision aiding?

Chapter 7

PREFERENCE, INDIFFERENCE, INCOMPARABILITY: BINARY RELATIONS AND BASIC STRUCTURES

SUMMARY

In the first section we present concepts that describe an actor Z's stated preference judgments when comparing two actions of A. In Section 7.1.1.1 we use examples to introduce the basic preference situations with which Z can be faced. We then state the axiom of limited comparability (Axiom 7.1.1), which serves as a point of departure from classic decision theory.

Based on the four binary relations of indifference (I), strict preference (P), weak preference (Q), and incomparability (R), we show in Section 7.1.2.2 that all of Z's preferences on A can be modeled by a basic system of preference relations, denoted BSPR, or, if necessary, by a consolidated system of preference relations, denoted CSPR. We discuss the two nontraditional relations Q and R in Section 7.1.2.3 and discuss transitivity of the relations I, P, Q, and R in Section 7.1.2.4.

We then investigate the situations and consolidated relations defined in Table 7.1.5 and the most noteworthy CSPR's that they generate. The one on which classical decision theory is based consists of two consolidated relations: $\sim$ (nonpreference) and $\succ$ (preference). In Section 7.1.3.2 we discuss how the axiom of limited comparability is replaced by the axiom of complete transitive comparability in this theory. After discussing the three consolidated relations J, K, and S, in Section 7.1.3.3, we introduce a final CSPR, one in which the outranking relation S plays a fundamental role. The section ends by examining relationships among the various situations and motivating certain choices of SPR models.

We use the first part of the second section to introduce graphical conventions used in the rest of the book. We also present a new example concerning a mayor's preferences. In the two subsections that follow, we present the principal structures associated with the most interesting SPR's and their functional representations. In Section 7.2.2 we look at those that exclude incomparability, and in Section 7.2.3 at those that acknowledge it. The following table synthesizes the principal structures studied in these two subsections in terms of the relations that constitute the SPR.

We touch upon the fundamental problem of comparing preference differences or exchanges (elements of $A \times A$) in Section 7.2.4.1 and examine the relationships between SPR's on A and on $A \times A$ in Section 7.2.4.2.

Readers put off by the terse nature of Section 2 should only skim its first two subsections.

Number of Relations	Properties Relating to Symmetry	Properties Relating to Transitivity	Other Properties	Name	Section	References
1	symmetric	transitive	–	equivalence class	7.2.2.1a	Fig. 7.2.6
		intransitive	–	(none)	–	–
	asymetric	transitive	–	complete order	7.2.2.1b	Res. 7.2.1, Fig. 7.2.8
		intransitive	–	intransitive tournament	7.2.2.1b	Fig. 7.2.7
	–	–	–	complete BSOR	7.2.2.1c	–
2	1 symmetric (T) and 1 asymmetric (V)	both transitive	–	complete preorder	7.2.2.2a	Fig. 7.2.9
		only V transitive	$V\,T\,V \subset V$ $V^2 \cap T^2 = \emptyset$	semi-order	7.2.2.2b	Fig. 7.2.10, Def. 7.2.1, Res. 7.2.3
			$V\,T\,V \subset V$	interval order	7.2.2.2c	Def. 7.2.2 (r 7.2.4)
			–	partial order	7.2.2.2c	Def. 7.2.2 (r 7.2.5)
3	1 symmetric (T), 2 asymmetric (V and W)	V transitive	yes	–	7.2.2.3a	(r 7.2.8)
			yes	pseudo-order	7.2.2.3b	Fig. 7.2.16, Def. 7.2.3, Res. 7.2.4
		T and V transitive	yes	directed semi-order	7.2.2.3c	Def. 7.2.4, Res. 7.2.5
	2 symmetric (R and T) 1 asymmetric (V)	T and V transitive	R is T-transitive	partial preorder	7.2.3.2	Def. 7.2.5

7.1 GENERAL COMMENTS ON PREFERENCE MODELING: BASIC CONCEPTS

7.1.1 Basic preference situations

Consider an actor Z who must furnish a preference judgment relative to two potential actions. Suppose that she has been informed of the relevant consequences[1] of the actions, even though the consequences may not be described (or even known) completely or precisely. We shall use the three examples introduced in Chapters 1 and 2 to illustrate the different situations that Z might face. We then propose an axiom that broadens the perspective of classical decision theory and forms the basis for a radically different approach for addressing many important issues that may arise.

7.1.1.1 Introductory examples

Family car example

In the anecdote presented at the end of Chapter 1, the father (Actor Z) has finally reached the point of comparing the four automobile models whose characteristics are listed in Table 7.1.1. He considers comfort, safety, and cost per mile very important, but aesthetics and maximum speed as minor factors. The sales price only becomes important as it approaches a fixed budget, B.

Table 7.1.1: The family car example

Automobile Model	Comfort	Safety	Cost per Mile Index	Aesthetics	Maximum Speed	Sales Price
a_1	Overall satisfactory, but size is slightly insufficient	Normal	0.39	Acceptable	85 mph	$0.87 \times B$
a_2	Particularly satisfactory and spacious	Normal	0.41	Elegant	90 mph	$0.95 \times B$
a_f	Acceptable but cramped	Normal	0.66	Very elegant	110 mph	$0.99 \times B$
a_n	Particularly satisfactory and spacious	Better than normal	0.40	Elegant	90 mph	Probably larger than B

[1] *We define this concept in Chapter 8, which can be read in parallel with Chapter 7.*

Given the data and Z's values, he clearly would attach a strong preference for a_1 or a_2 over a_f. We shall say that a_1 (or a_2) is **strictly preferred** to a_f.

On the other hand, the comparison between a_1 and a_2 is not as straightforward. Even if a_1 is significantly less expensive than a_2, a_2 would still offer a slight advantage on the basis of comfort and aesthetics. (Assume that the difference in cost per mile can be considered insignificant.) The father does not consider the advantages overwhelming, however, and he does not accept the statement that a_2 is strictly preferred to a_1. Given the relatively low importance that he attaches to the criterion of "aesthetics," a_1's sufficient level of comfort, and the difference in sales price, he is inclined to have a slight preference for a_1 over a_2. Still, this preference is far from being as strong as that which he has for a_1 over a_f. He might place a_1 and a_2 at the same level in declaring that he is indifferent between these two models; but this would imply that he not only has no preference in favor of a_2 over a_1, but that he likewise has no preference in favor of a_1 over a_2, which is not the case. We characterize this situation by saying that a_1 is **weakly preferred** to a_2. Specifically, this would indicate that the actor Z would hesitate between a_1 strictly preferred to a_2 and a_1 indifferent to a_2.

Finally, not knowing the purchase price of a_n makes it dangerous to compare the other cars with this model. Let us denote the unknown purchase price of a_n by $k \times B$. If k proves to be on the order of 0.9, Z would strictly prefer a_n to a_1 or a_2. If, however, k reaches or exceeds 1.2, Z would exhibit the opposite preference. He considers the value k_1 of k which would render him indifferent between a_1 and a_n very difficult to determine and, moreover, of little interest. Under these conditions, we will speak of a **situation of incomparability** to characterize the lack of a preference judgment between a_n and either a_1 or a_2.

Highway location example

Here we are interested in locating a highway section to be included in an already approved project. This section will serve as a bypass to a large residential area and, therefore, be located near a heavily populated region. There also exists a forested area in the region, and preserving the forest is considered important. The negative impacts on the populated and forested areas could be reduced by circumventing these areas, but this would lengthen the section. Alternatively, large parts of the section could be built below grade or as tunnels. Both approaches would increase the cost of the project, however, and decrease the level of service offered to the road users. Four locations are proposed for consideration by a committee Z (see Table 7.1.2).

The committee considers the numbers representing the consequences of a_1 and a_2 sufficiently clear and positive to justify the equivalence of the two locations – that is, the committee Z is said to be in a situation of **indifference**. Specifically, in comparing a_1 and a_2 according to the first two criteria representing the consequences, then the next two, and finally the last two, Z considers the differences to be small and to balance out.

When comparing the numbers of a_1 and a_3, Z feels that:

- the two criteria relating to the impact on the population balance out, or perhaps can be considered slightly in favor of a_3;
- the two criteria relating to the deforestation are slightly in favor of a_3;
- the last two criteria balance out, or are slightly in favor of a_1.

Table 7.1.2: The highway location example

Location	Thousands of people inhabiting the zone with greatest noise level	Thousands of people requiring relocation	% forest destroyed	% forest damaged but not destroyed	Added length in km	Financial cost in millions of francs
a_1	7.5	53	12.5	27.5	0	511
a_2	7.0	56	13.0	25.0	0.5	502
a_3	6.6	58	13.5	20.0	1.5	490
a_4	4.0	23	0.7	0	5.0	1104

Given the importance that Z ascribes to the first four criteria, the committee agrees that a_1 cannot be strictly preferred to a_3, but it is not ready to state strict preference for a_3. The committee refuses to proclaim indifference between the two alignments, however, wishing instead to manifest a slight preference for a_3. In this case, we find once again a situation of weak preference.

The differences in consequences when comparing a_3 and a_4 are much greater than those when comparing a_1 and a_3. The issue is whether the improved impacts on the inhabitants and the forested area associated with a_4 justify the increased travel time associated with the extra 3.5 km of highway and, especially, the 225 percent increase in project costs. Some members feel that it does; others feel that it does not. The committee cannot reach an agreement, and we shall say that Z finds itself in a situation of incomparability when considering a_3 and a_4.

Loan application example

Consider the case of an employee responsible for approving a certain type of "loan." Suppose that her supervisors (Actor Z) want her to decide on approving or rejecting the loan on the basis of five ratios that are easily calculated from the application information. Previous studies have resulted in combinations of values for these five ratios that would lead Z to classify a loan as definitely acceptable or definitely unacceptable (see Table 7.1.3).

Specifically, Z would automatically accept any application with a set of ratios at least as good as b_1's or b_2's and automatically reject any application with a set of ratios as bad as or worse than c_1's or c_2's. These combinations of ratio values that characterize the "certainly acceptable" and "certainly unacceptable" were established by referring to categories of individuals applying for loans. Even though it is difficult to compare acceptable benchmarks b_1 and b_2, or unacceptable benchmarks c_1 and c_2, there is no reason to do so. On the other hand, it is clear that b_1 is strictly preferred to c_1 and that b_2 is strictly preferred to c_2.

Table 7.1.3: The loan application example

		Ratio 1	Ratio 2	Ratio 3	Ratio 4	Ratio 5
Definitely acceptable limits	b_1	0.70	0.70	0.50	0.50	0.50
	b_2	0.50	0.50	0.70	0.70	0.70
	...	...	...	...	...	...
Definitely unacceptable limits	c_1	0.50	0.50	0.50	0.50	0.50
	c_2	0.50	0.50	0.30	0.70	0.50
	...	...	...	...	...	...
Example loan applications	a_1	0.72	0.73	0.54	0.50	0.50
	a_2	0.50	0.50	0.60	0.70	0.70
	a_3	0.72	0.73	0.30	0.70	0.55
	a_4	0.60	0.60	0.30	0.70	0.24
	a_5	0.60	0.60	0.60	0.50	0.12
	a_6	0.50	0.50	0.40	0.58	0.50
	...	...	...	...	...	...

The employee must make her decision on real applications, such as a_1, a_2, ... She is to do so in such a way that the decisions conform with what she believes to be Z's preferences. She must also keep in mind that the ratios have been calculated from imperfect data. If she were to ignore this point, she would conclude that a_1 is strictly better than b_1, even though this superiority may be a result of "errors" or "noise" in data collection or processing. She may, therefore, want to say that a_1 is equivalent to b_1, or at most slightly preferred to it.

7.1.1.2 Basic situations and the axiom of limited comparability

These examples illustrate that comparing any two potential actions a and a′ puts the actor Z in one of the following four basic situations:

- **situation of indifference**, in which there is only one possibility: a and a′ are of equal value;
- **situation of strict preference**, in which there are two possibilities: either a is strictly preferred to a′, or a′ is strictly preferred to a;
- **situation of weak preference**, in which there are two possibilities: a is weakly preferred to a′, or a′ is weakly preferred to a.
- **situation of incomparability**, in which there is only one possibility: a and a′ are incomparable.

We define each of these situations in Table 7.1.4. Note that **any two situations are mutually exclusive**; the actor Z can, nevertheless, retain two or three of these situations relative to a single pair a, a′ when it seems to her that it is impossible, not useful, or too early to refine her analysis or judgment any further. On the other hand, as we shall discuss later, it does not seem overly restrictive to propose that these four situations exhaust the set of possible situations.

In practical decision aiding, it would be rare for actor Z to be the one to express which situation she would accept for the different pairs of actions. More often, the analyst considers the available information and the value system he postulates for the actor Z in proposing one of the four basic situations or, if necessary, two or three among them. What are expressed, then, are **Z's preferences as perceived by the analyst**. In this case, we will say that **the analyst judges in Z's name**.

We are now ready to state the axiom on which the rest of this book is largely based.

AXIOM 7.1.1 Axiom of limited comparability: The four conflicting basic situations of indifference, strict preference, weak preference, and incomparability (defined in Table 7.1.4) are useful in establishing a realistic representation of the actor Z's preferences; whatever the actions considered, the point of view taken to compare them, and the information available, Z or the analyst judging in Z's name can develop a satisfactory model that develops or documents[2] *Z's preferences by assigning one, or a grouping of two or three of these four basic situations, to any pair of actions.*

Classical decision theory introduces only two basic situations: indifference and strict preference. The axiom corresponding to that theory is, therefore, more restrictive (see Section 7.1.3). The situations of incomparability and weak preference are either treated as if they do not exist or are combined into indifference and strict preference, respectively.[3] Excluding these two situations presents serious difficulties for decision aiding, however. We shall return to these difficulties on several occasions.

At this point, we wish only to emphasize that the analyst judging in Z's name (as well as the actor Z herself) has several reasons to try to avoid the "indifference or strict preference" dilemma when having to compare two actions a and a′. One may:

a) *not be able to decide*: The data could be subjective or have been collected somewhat hastily, possibly making it inappropriate for a categorical judgment that would allow only indifference or strict preference. To transform a weak preference into a strict preference based on some indices or to argue that lack of data should lead to indifference seems somewhat arbitrary and incoherent.

---

[2] *We shall discuss this distinction in Section 10.3 when presenting the difference between descriptive ("documenting") and constructive ("developing") approaches.*

[3] *See Arrow (1963, p. 13) or Fishburn (1970, p. 12) for example. See also Roubens and Vincke (1985).*

b) not know how to decide: The analyst may have no feeling for the decision maker's preferences for certain pairs of actions, either because the decision maker is absent or inaccessible (head of state, president of a large firm), or because she is a vague entity (public opinion) or a group (committee) whose preferences are ill-defined and partially contradictory.

c) not wish to decide: Since it involves weighing the pros and cons of a versus those of a′ without neglecting the common elements of a and a′, proposing either strict preference or indifference implies sufficient information of Z's values or voluntaristic (or goal-oriented) hypotheses to resolve the conflict. Obtaining the necessary information might take too long or cost too much, or the analyst might wish to wait until later in the study to introduce any voluntaristic hypotheses. In any case, the analyst may not wish to commit at this point and opt for weak preference or incomparability.

7.1.2 Modeling with binary relations: System of preference relations

A binary relation links two objects and describes the presence or absence of a certain property. This concept is well suited to modeling the situations introduced above. Each can be considered a "property" that a pair[4] of actions a, a′ either does or does not possess (see Table 7.1.4).

We next discuss the notation and terminology of the right hand column of Table 7.1.4, then in light of the axiom of limited comparability, use them to characterize the two principal types of relational systems by which preferences of Z on A can be modeled. Finally, the last two sub-subsections treat various aspects concerning the binary relations.

7.1.2.1 Notation and terminology

The situation of indifference allows only one possibility between any two actions a and a′ and can be denoted by any of the following:

a I a′; a′ I a; a I a′ is true; a′ I a is true.

The order in which a and a′ appear in the notation is unimportant, and the relation is said to be **symmetric**. It is also **reflexive** – i.e., when a′ is identical to a, the relation I holds. Finally, we must be able to denote the case in which indifference does not hold between a and a′. This can be written either as:

not a I a′;

or as:

a I a′ is false.

[4] *By using the word "pair" instead of "couple", we emphasize the fact that the two actions are considered independently of the order in which they are presented. That is, a, a′ and a′, a are identical pairs but distinct couples.*

Table 7.1.4: The four basic preference situations for comparing two potential actions

Situation	Definition	Binary Relations (properties)
Indifference	Corresponds to the existence of clear and positive reasons that justify equivalence between the two actions.	I: reflexive and symmetric relation
Strict Preference	Corresponds to the existence of clear and positive reasons that justify significant preference in favor of one (identified) of the two actions.	P: asymmetric (nonreflexive) relation
Weak Preference	Corresponds to the existence of clear and positive reasons that invalidate strict preference in favor of one (identified) of the two actions but that are insufficient to deduce either strict preference in favor of the other action or indifference between the two actions, thereby not allowing either of the two preceding situations to be distinguished as appropriate.	Q: asymmetric (nonreflexive) relation
Incomparability	Corresponds to an absence of clear and positive reasons that justify any of the three preceding relations.	R: symmetric (nonreflexive) relation

In the highway location example (see Table 7.1.2), we can write a_1 I a_2; not a_3 I a_1, or a_1 I a_3 is false.

In the case of strict preference we have seen that there are two possibilities:

- a strictly preferred to a′, which we write either: a P a′; or: a P a′ is true;
- a′ strictly preferred to a, which we write either: a′ P a; or: a′ P a is true.

In the family car example (see Table 7.1.1), we can write a_1 P a_f or a_f P a_1 is false. Since the order in which a and a′ appear in the relations is important, P is not symmetric. In addition:

$$a \text{ P } a' \text{ implies not } a' \text{ P } a,$$

which means that P is asymmetric. It follows then that a P a is false for any action a; i.e., P is nonreflexive.

Replacing P by Q in the above, the same properties and notation hold for the case of weak preference. In the highway alignment example, we can write, not a_1 Q a_3, but a_3 Q a_1. In the family car example, we can write, a_1 Q a_2 is true. As for the case of incomparability, it should be clear that the relation R is symmetric and nonreflexive. In the family car example, a_1 R a_n; and in the loan application example, b_1 R b_2, and c_1 R c_2, but not c_1 R b_1.

Using this notation, we can summarize the father's preferences (see Table 7.1.1) as:

$$a_l \; Q \; a_2; \; a_l \; P \; a_f; \; a_2 \; P \; a_f;$$
$$a_n \; R \; a_l; \; a_n \; R \; a_2; \; a_n \; P \text{ (or R) } a_f. \qquad \text{(r 7.1.1)}$$

7.1.2.2 Systems of preference relations and the axiom of limited comparability

Let us now see how the above binary relations can be used with the axiom of limited comparability to model the set of preferences that an actor Z has on the set A.

One way to proceed is by allowing one (exhaustive condition) and only one (mutually exclusive condition) of the four fundamental relations to hold for each pair of actions a and a′. This is equivalent to considering one and only one of the following six statements as true: a I a′; a P a′; a′ P a; a Q a′; a′ Q a; a R a′.

Two of the four binary relations that we have defined, P and Q, are asymmetric. The two others, I and R, are symmetric. Also, one can always state a I a, for all $a \in A$. When Z, or the analyst judging in her name, can assign the four exhaustive and mutually exclusive binary relations I, P, Q, R on A without ambiguity, we shall say that the model representing Z's preferences is a basic system of preference relations (BSPR).

DEFINITION 7.1.1: We shall say that the four binary relations I, P, Q, R defined over a set of potential actions A form a ***basic system of preference relations (BSPR)*** *for an actor Z over A if:*

i) they can represent Z's preferences with respect to the actions in A according to the definitions and properties of Table 7.1.4;

ii) they are exhaustive: for any pair of actions, at least one of the relations holds;

iii) they are mutually exclusive: for any pair of actions, at most one of the relations holds.

As an example, the information presented in Table 7.1.2 and the discussion of the highway location example in Section 7.1.1.1 form the following basic system of preference relations for committee Z over $A = \{a_1, a_2, a_3, a_4\}$:

$$a_1 \; I \; a_2; \; a_3 \; Q \; a_1; \; a_4 \; R \; a_1; \; a_2 \; I \; a_3; \; a_4 \; R \; a_2; \; a_4 \; R \; a_3. \qquad \text{(r 7.1.2)}$$

In the loan application example, we can consider the set $A = \{b_1, b_2, c_1, c_2\}$ (see Table 7.1.3) and the two relations defined over this set to say:

$$b_1 \; R \; b_2; \; b_1 \; P \; c_1; \; b_1 \; P \; c_2; \; b_2 \; P \; c_1; \; b_2 \; P \; c_2; \; c_1 \; R \; c_2. \qquad \text{(r 7.1.3)}$$

For the employee handling the applications, the system (r 7.1.3) is a BSPR on A. The fact that Q holds for no pair of actions will be denoted from now on as: $Q = \varnothing$. Although there is no distinct pair of actions satisfying the relation I, $I \neq \varnothing$, since a I a, $\forall a \in A$. Because of this property we did not state it explicitly in (r 7.1.3) and shall not state it except where needed for clarity. We shall express the fact that Q and only Q is empty in this example by saying that the BSPR has the form (I, P, R).

Finally, the system (r 7.1.1) (see Section 7.1.2.1) does not define a BSPR for the father, since the mutually exclusive property does not hold – the system does not commit to either a_n P a_f or a_n R a_f. Even though one pair of actions corresponds to two basic relations, system (r 7.1.1) still respects the axiom of limited comparability. We have already mentioned (see Section 7.1.1.2) that for various reasons (ambiguity, reluctance, lack of knowledge, ...), two or three of the basic relations can represent preferences for a single pair of actions, even though the relations are mutually exclusive. This leads us to introduce a second type of relational system for modeling Z's preferences on A.

This second system allows more than one of the basic situations to hold for each pair of actions. The fact that two or more situations are considered possible for a given pair of actions is due to the fact that Z, or the analyst representing her, cannot, does not wish to, or does not know how to decide upon the appropriate situation. It follows, then, that when two or three of the six statements (a I a′; a P a′; a′ P a; a Q a′; a′ Q a; a R a′) are considered possible, it is not because they are considered to hold simultaneously, but because it is thought impossible, premature, or not useful to determine which of them does, in fact, hold.

Note that it does not make sense for a P a′ and a′ P a to appear simultaneously in a group of preference statements; neither would it make sense for a Q a′ and a′ Q a to appear. Note, also, that our preceding discussion says nothing of how strict preference (P) and weak preference (Q) can appear together in a group of preference statements. It should be clear, however, that hesitating between a P a′ and a Q a′ would seem consistent, while hesitating between a P a′ and a′ Q a would not. Similarly, one would only very rarely expect a hesitation between a Q a′ and a R a′. Thus, among all the possible groupings of preference statements, it appears necessary to distinguish those of most interest. In Section 7.1.3 we propose five such groupings. These can be found in Table 7.1.5.

Representing preferences by using at least one of these five relations, as opposed to using one or more of the four basic relations, leads to a model which we call a consolidated system of preference relations (CSPR).

DEFINITION 7.1.2: We say that the nine binary relations I, R, ~, P, Q, ≻, J, K, S defined over a set of possible actions A, form a ***consolidated system of preference relations (CSPR)*** *for an actor Z on A if:*

i) they can represent Z's preferences with respect to the actions in A according to the definitions and properties of Tables 7.1.4 and 7.1.5;

ii) they are exhaustive: for any pair of actions, at least one the relations holds;

iii) they are mutually exclusive[5]: for any pair of actions, two distinct relations cannot hold;

[5] *See comment c) below.*

iv) at least one of the five relations ~, ≻, J, K, S is not empty.

This definition leads to the following comments:

a) In general, a CSPR will not include all nine relations; the different forms that a CSPR may take in practice will be discussed in Section 7.2.

b) The notation and definitions of the relations (Tables 7.1.4 and 7.1.5) require I, R, ~ to be symmetric and P, Q, ≻ to be asymmetric; J, K, S, on the other hand, are neither symmetric nor asymmetric, as they combine symmetric and asymmetric binary relations (see Section 7.1.3.3).

c) Condition iii) requires that, if H_1 and H_2 are two distinct relations among the nine, then:

- [a H_1 b and b H_2 a] is excluded: this condition follows from the definitions of Tables 7.1.4 and 7.1.5 except in pathological cases that are unimportant in practice.
- [a H_1 b and a H_2 b] is excluded: unlike the preceding condition this might seem restrictive, since either [a P b and a ≻ b], [a ≻ b and a J b], or [a I b and a S b] represent coherent situations.

But in all of these types of coherent situations, it is possible to use one relation instead of two. For example, in the above:

[a P b and a ≻ b] can be replaced by a P b;
[a ≻ b and a J b] can be replaced by a Q b;
[a I b and a S b] can be replaced by a I b.

Condition iii) is, therefore, introduced so that only those systems of relations that are neither redundant nor contradicted by the definition of the nine relations will be called a CSPR.

d) The developments of Section 7.1.3 will help the reader better understand the importance and interest of these consolidated systems of relations. Although they are less refined than the basic systems of relations from which they are derived, the CSPR's are easier to develop.

e) A relational system satisfying all but Condition iv of Definition 7.1.2 is a BSPR.

The need for a general term describing the two types of preference models introduced leads to the following definition.

*DEFINITION 7.1.3: A **system of preference relations (SPR)** is a model of actor Z's preferences over a set of potential actions A that is either a BSPR or a CSPR.*

Finally, we add that SPR's are not the only models compatible with the axiom of limited comparability. Still, although using only the five binary relations of Table 7.1.5 may represent a theoretical restriction, we shall argue in Section 7.1.3.4 that the restriction is justified in practice.

7.1.2.3 Comments on incomparability and weak preference

The motivation for introducing (see Roy, 1973 and 1977) the binary incomparability and weak preference relations to complement the more traditional indifference and strict preference relations should be evident from the discussion already presented. We offer a few further comments at this point, however.

Some would claim that any decision essentially eliminates situations of incomparability. This claim misses the point, however. Specifically, one should remember that a R a′ represents a refusal to evaluate a relative to a′ at the **preference modeling level**.

This refusal cannot always be assumed (as is done in classical theory; see Section 7.1.3.2) to mean that Z is indifferent between a and a′. We shall see later that one will need to resort (at least temporarily) to situations of incomparability when confronted with multiple criteria.

As for weak preference, stating that a Q a′ indicates that Z or the analyst modeling Z's judgments definitely feels that not a′ P a, but that under the present conditions she cannot state whether a P a′ or a I a′. This intermediate situation between I and P might even be quantified by using a fuzzy number (see Dubois and Prade, 1980 and Perny and Roy, 1992) or credibility index of a proposition to denote the relative distance between the extremes of indifference and strict preference.

Although the motivation for the weak preference relation should be apparent, one might wonder if it would be better to introduce it as a consolidation of the relations P and I, rather than to consider it on the same plane as the other three basic relations. Technically, we could have done so. We chose not to, however, since weak preference can be the result of an irreducible situation, one that conceptually deserves to be placed at the same level as indifference or strict preference. Moreover, treating weak preference as an additional consolidation of relations would either weaken our modeling capabilities by prohibiting certain subtleties or increase the complexity of various definitions like the relations J and S.

7.1.2.4 Comments on the transitivity of the basic binary relations

The only noteworthy property that the four relations I, P, Q, R are defined to possess is symmetry or asymmetry. Nevertheless, many approaches require I and P to be transitive.

A binary relation H is **transitive** if and only if it is impossible to find a, a′, a″ (different or not) such that:

a H a′, a′ H a″, not a H a″.

It should be clear that there is no reason for the incomparability relation R to be transitive. For example, in the BSPR (r 7.1.2), one can see that a_1 R a_4 and a_4 R a_2, but a_1 I a_2 and not a_1 R a_2; and also that a_3 R a_4 and a_4 R a_1, but a_3 Q a_1.

Consider now the indifference relation I, which is often postulated to be transitive. Referring to a well known example (see Luce, 1956), one might be indifferent between a cup of coffee without sugar and the same cup with a half spoon of sugar, and be indifferent between the cup with a half spoon of sugar and the same cup with a full spoon of sugar, yet prefer the cup without sugar to that with a full spoon (or vice-versa depending on whether or not one likes sugar in coffee). Note also that in (r 7.1.2), we have a_3 I a_2, a_2 I a_1, but a_3 Q a_1. **We shall not, therefore, require I to be transitive. Neither shall we require transitivity of Q**. It will be quite possible to have a Q a′, a′ Q a″, and a P a″.

The one relation that we might want to require to be transitive is that of strict preference. Even though P is commonly held to be transitive, it would not be if a P a′, a′ P a″, and a R a″. This might be the case, for example, if a and a″ each had enough features in common with a′ to allow the strict preferences, but had so few in common with each other that the analyst had to delay in pronouncing a judgment.

Following the same line of thought, recall that a relation P based on majority preference (a concept that would be hard to consider irrational) is not necessarily transitive. Indeed, as Condorcet (1785) showed, in a group of individuals each with transitive strict preferences, the majority can prefer a to b, b to c, and c to a. (The composition of the majorities would, or course, be different in each of the three cases.) One can find in Schärlig (1985, Chapters 1 to 8) other versions of this "Condorcet paradox." The reader can also find more on the subject in Weinstein (1968), Tversky (1969), Schwartz (1972), and Roubens and Vincke (1985).

In short, we shall not necessarily require P to be transitive.

7.1.3 Consolidated situations and associated binary relations

7.1.3.1 General comments

As explained in Section 7.1.2.2, it is often useful to consider a few binary relations that group together two or three of the basic relations I, P, Q, and R, when modeling preferences. Among all the groupings that are possible, at most five have any real importance. We shall talk of **consolidated systems of preference relations**, CSPR's (Table 7.1.5), and call the binary relations associated with them **consolidated relations**.

Before commenting on these definitions, we emphasize the following two points:

- A CSPR is not only used in the analysis to avoid pointless or premature efforts in discovering the most appropriate situation for a condition. It might also be used to synthesize basic situations that are well specified. In the latter case, being more explicit would, of course, decrease the ambiguity as to the origin of the consolidated situations.

- A CSPR should not be considered to be motivated by the need to approximate a poorly understood reality. That is, it may be tempting to postulate the existence of "true preferences" that can only be modeled partially and with error. Such an idea would force us to distinguish between "true preferences" and "modeled preferences." We consider the distinction confusing, however, since the concept of pre-existing true preferences cannot be justified in many cases. Preferences are at least partially ill-defined, delicate, and conflicting by nature. As we shall see, the CSPR's that are used as decision aids are more often "constructs," used as a basis for discussion or deduction, than representations of some reality that is independent of the analyst.

Classical theory[6] (i.e., classical decision theory) considers only two consolidated situations, regardless of the basic situations that they consolidate. Considering these two situations and the basis of classical theory leads us to define an important special case of a CSPR in the next subsection. In Section 7.1.3.3 we present the last three consolidated situations of Table 7.1.5, introduce a special case of a CSPR where the outranking relation plays an important role, investigate relations among the various situations, and explain some of the subtler choices that were made.

7.1.3.2 Preference and nonpreference: Perfect system of preference relations

By definition, the nonpreference relation ($\sim$) is symmetric and reflexive, whereas the preference relation ($\succ$) is asymmetric and, therefore, nonreflexive. Because the two situations are complementary, it follows that, for any pair of actions, one and only one of the following statements is true:

$$a \sim a'; \; a \succ a'; \; a' \succ a.$$

Since the indifference (I) and incomparability (R) relations are not necessarily transitive, there is no reason for the nonpreference relation to be transitive. Given the discussion in Section 7.1.2.3, transitivity of the preference relation ($\succ$) would also be hard to justify.

Classical theory is based only on the two relations $\sim$ and $\succ$, and since it does not explicitly consider incomparability and weak preference, it associates $\sim$ and $\succ$ with indifference and strict preference, respectively. Moreover, $\sim$ and $\succ$ are automatically assumed to be transitive. Classical theory is, therefore, based on an axiom that reduces the number of basic situations from four to two (implying perfect comparability) and

---

[6] *From now on, we shall use "classical theory" for classical decision theory; see, e.g. Arrow (1963), Fishburn (1970), Raiffa (1970).*

Table 7.1.5: Consolidated relations and situations for modeling preferences relative to two potential actions a and a′

Situation	Definition	Binary relation (properties)
Nonpreference	Corresponds to an absence of clear and positive reasons that justify strict or weak preference in favor of either of the two actions and thus consolidates situations of indifference and incomparability without being able to differentiate between them.	~[7]: a ~ a′ ⇔[8] a I a′ or a R a′
Preference	Corresponds to the existence of clear and positive reasons that justify strict or weak preference in favor of one (identified) of the two actions and thus consolidates situations of strict and weak preference without being able to differentiate between them.	≻[1]: a ≻ a′ ⇔ a P a′ or a Q a′
J-preference	Corresponds to the existence of clear and positive reasons that justify weak preference, no matter how weak, in favor of one (identified) of the two actions or, at the limit, indifference between the two actions, but with no significant division established between the situations of weak preference and indifference.	J: a J a′ ⇒[9] a Q a′ or a I a′
K-preference	Corresponds to the existence of clear and positive reasons that justify strict preference in favor of one (identified) of the two actions or incomparability between the two actions, but with no significant division established between the situations of strict preference and incomparability.	K: a K a′ ⇒ a P a′ or a R a′
Outranking	Corresponds to the existence of clear and positive reasons that justify either preference or J-preference in favor of one (identified) of the two actions but with no significant division being established among the situations of strict preference, weak preference and indifference.	S: a S a′ ⇒ a P a′ or a Q a′ or a I a′

[7] *As it is formally defined, this relation can violate condition iii) of Definition 7.1.2 when the CSPR uses basic relations. Given comment c) following the definition, one can easily overcome this difficulty by adding the condition that only the basic relation holds for the pairs in question.*

[8] ⇔ *is read, "if and only if."*

[9] ⇒ *is read, "only if".*

excludes a certain number of other situations of interest through the transitivity imposed. That is, the axiom is doubly restrictive compared to axiom 7.1.1 and can be stated as follows:

AXIOM 7.1.2 Axiom of perfect, transitive comparability: The two conflicting situations of nonpreference and preference defined in Table 7.1.5 are sufficient to form a realistic representation of actor Z's preferences; whatever the actions considered, the point of view taken to compare them, and the information available, Z or the analyst judging in Z's name can develop a satisfactory model that develops or documents Z's preferences by assigning exactly one of these two situations to any pair of actions in such a way that they are both transitive.

Note that this axiom does not necessarily exclude incomparability. Indeed, ~ could cover situations both of indifference and incomparability. The axiom does not explicitly separate situations of incomparability from those of indifference, however, and implies that they can be treated identically.

To accept Axiom 7.1.2 as a basis for preference modeling implies accepting very special cases of CSPR's and BSPR's that we shall call perfect systems of preference relations (PSPR's).

DEFINITION 7.1.4: A ***perfect system of preference relations (PSPR)*** *is a system of two transitive preference relations, formed by combining either I or ~ with either P or ≻.*

A PSPR can, therefore, be thought of as either a BSPR of the form (I, P) or a CSPR of the form (~, ≻), (~, P), or (I, ≻).

We call attention to the fact that postulating the transitivity of an asymmetric relation T (e.g., ≻ or P) does not imply the transitivity of the complementary relation $\hat{T}$, defined as:

$$a \ \hat{T} \ a' \Leftrightarrow \text{not } a \ T \ a' \text{ and not } a' \ T \ a.$$

7.1.3.3 J-preference, K-preference, basic system of outranking relations

Let J^1 be the binary relation defined by:

$$a \ J^1 \ a' \text{ if and only if } a \ Q \ a' \text{ or } a \ I \ a'. \qquad \text{(r 7.1.4)}$$

For some pairs of actions, J^1 is symmetric, while for other pairs, it is asymmetric. The asymmetric cases are those corresponding to weak preference, whereas the symmetric cases correspond to those of indifference. Since the relation is "if and only if," knowing both whether or not J^1 holds for the couple (a, a′) and whether or not it holds for the couple (a′, a), it is possible to recover the basic relations Q and I. This is not the case for the relation J, however, since it is an "only if" relation (see Table 7.1.5).

More formally, from the relations of Table 7.1.5 one can derive the following relations:

$$\begin{aligned} &a\ J\ a' \text{ and } \quad a'\ J\ a \Rightarrow \quad a\ I\ a';\\ &a\ J\ a' \text{ and not } a'\ J\ a \Rightarrow \text{either } a\ I\ a',\\ &\qquad\qquad\qquad\qquad\qquad \text{or} \quad a\ Q\ a'. \end{aligned} \tag{r 7.1.5}$$

The father in the family car example who does not have the time to refine his judgment relative to a_1 and a_2 (see Table 7.1.1) can simply propose a_1 J a_2 and not a_2 J a_1. This allows him to leave both the options of indifference and weak preference open while still indicating that if he had to decide on weak preference, it would be in the form of a_1 Q a_2.

In the same way, knowing only the relation K^1 defined as:

$$a\ K^1\ a' \text{ if and only if } a\ P\ a' \text{ or } a\ R\ a'; \tag{r 7.1.6}$$

for all couples of actions is enough to recover the basic relations P and R. For reasons parallel to those given above, this would not be the case with K-preference.

We note that it is possible to build a CSPR only with relations J and K. This is the primary reason that K-preference might appear in the modeling process. It might also be used in those cases where one would opt for strictly preferring one of the actions if it was **absolutely** necessary to compare them. Therefore, in the family car example, a_1 K a_n and not a_n K a_1 represents the father's hesitation to buy a_n because of its price. Another illustration of K is provided in (r 7.1.1).

The last relation defined in Table 7.1.5 has a particularly simple interpretation and, unlike the K-preference relation, is often of great interest. **We say that a outranks a′ if a is considered to be at least as good as a′**. Thus in the case of the loan applications, Z can state a_1 S b_1 (see Table 7.1.3) and not have to decide whether this means that a_1 is strictly preferred to, weakly preferred to, or indifferent to b_1. Notice again that:

$$\begin{aligned} &a\ S\ a' \text{ and } \quad a'\ S\ a \Rightarrow \quad a\ I\ a';\\ &a\ S\ a' \text{ and not } a'\ S\ a \Rightarrow \text{either } a\ P\ a',\\ &\qquad\qquad\qquad\qquad\qquad \text{or} \quad a\ Q\ a',\\ &\qquad\qquad\qquad\qquad\qquad \text{or} \quad a\ I\ a'. \end{aligned} \tag{r 7.1.7}$$

In the following chapters, we shall pay special attention to certain CSPR's that use the relation S. We, therefore, propose the following definition:

*DEFINITION 7.1.5: A **basic system of outranking relations (BSOR)** is a consolidated system of preference relations in which S is non-empty and which is:*
- *either reduced to S: the BSOR is then said to be complete or total;*
- *or of the form (S, R), (S, ~), or (S, ~, R): the BSOR is then said to be incomplete or partial.*

Note that in a BSOR of the form (S, R), we always have a S a, $\forall$ a $\in$ A. More generally, it follows from (r 7.1.7) that the symmetric part of S can always be considered to represent indifference situations. On the other hand, it is generally incorrect to consider the asymmetric part of S as representing situations of preference and ruling out situations of indifference.

7.1.3.4 Links among these and other relations

Figure 7.1 illustrates the links that exist among the different relations defined in Tables 7.1.4 and 7.1.5. We emphasize that each of the five consolidated relations is less rich in information than the system of basic relations that it consolidates. Constructing these consolidated relations requires less effort on the part of the analyst, since such a preference model does not distinguish among the basic situations that they consolidate.

Figure 7.1: Illustration of the links between consolidated and basic relations (the dashed lines correspond to groupings that are not equivalent to a simple union of the components)

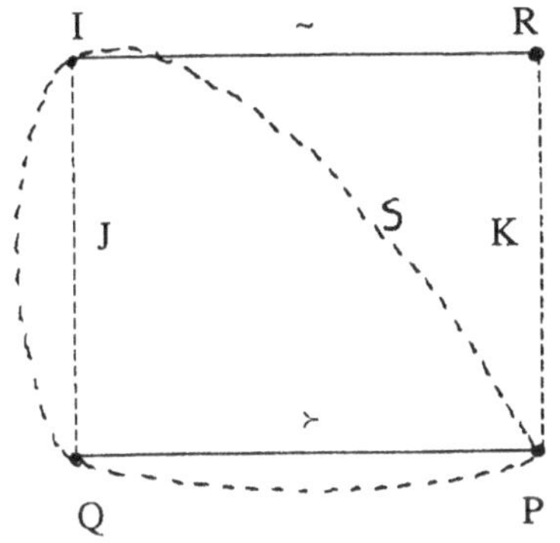

Figure 7.1 shows that Q and R remain unconsolidated. We could consolidate these two basic relations in a fashion similar to that in which K consolidates P and R, but there is little practical motivation for doing so. Similarly, consolidating I and P would be redundant with S (and not with Q, which is not a consolidation).

Logically, one could have considered two basic situations, different from the four already presented, to represent hesitations between indifference and incomparability and between incomparability and strict preference. Let us denote the binary relations that would model these hesitations M and N, respectively. With this convention, ~ would be defined by the triplet I, M, R; and K would be defined by the triplet R, N, P. Expanding the set of basic relations from four to six elements in this way has little value for realistic applications, however, while complicating the models and notation.

To conclude this section, we note that attempting to model an actor Z's preferences over a set A at a given phase of the investigation may lead the analyst to consider several SPR's. First of all, he must decide on the type of model to use: Is a BSOR sufficient? Would an enriched CSPR be preferable? Is it possible or necessary to develop a BSPR? Next, once the type of preference model is fixed, several different but nonconflicting SPR's could be built. Indeed, there are no general conditions to dictate the choice of one situation over another (see Tables 7.1.4 and 7.1.5). Consider the case of strict preference versus outranking, for example. Whether P or S is eventually used in the SPR will depend, in large part, on how convincing the analyst considers the arguments for one or the other. Even if general conditions could be established, certain relations such as J and S would still not be defined univocally, since their symmetric part does not necessarily reflect all the indifferent situations.

Classical (we shall even say trivial) binary relations imply that the statement a H a′ must either be true or false for a pair of actions a and a′ and a certain relation H when, in reality, such a conclusion cannot be reached. As a result, when faced with such a possibility, the analyst may be forced to conclude arbitrarily that a H a′ is either true or false. We shall see later that the concept of a fuzzy binary relation can reduce the arbitrary part that results from this type of difficulty.

7.2 PRINCIPAL STRUCTURES AND FUNCTIONAL RELATIONS[10]

The basic concepts presented above can be used to develop or understand a model that incorporates a representation of what are, can be, or might become actor Z's preferences. Using these concepts correctly, however, requires some understanding of certain structures and common problems. We discuss these before proceeding to the more concrete and operational aspects of preference modeling in subsequent chapters.

To make the discussion less abstract and illustrate the major systems of preference relations defined above, the first subsection is devoted to graphical representations, which will also be useful for the remainder of the book. In this subsection, we also present a new example that will be used in the same way as the family car, highway location, and loan application examples.

The following two subsections present the principal structures associated with the most interesting systems of preference relations. We look first at those that exclude (or obscure) incomparability, then at those that allow it. The last subsection offers a preview of the subtle problem of comparing and evaluating preference differences.

As will become clear after reading Chapter 9, all three subsections deal with representing preferences in such a way as to illustrate the concept of criterion.

---

[10] *Readers not interested in rigorous descriptions and definitions can skip ahead to Chapter 8 and come back to this section only when it is suggested to do so (mostly in Chapter 9). Nevertheless, we recommend skimming the first two subsections at this point.*

7.2.1 Graphical representations and an example system of preference relations

7.2.1.1 Graph theory: Notation

a) General notation

Let H be a binary relation (I, P, $\succ$, S, ...) defined on a set A assumed to be finite. It is always possible to represent H by a diagram, consisting of points and lines, called a **graph**. The points, called **vertices** of the graph, identify the elements of A. The lines connect pairs of vertices (elements) for which H is true. More specifically, if H is a symmetric relation, a line called an **undirected edge** connects two vertices a and a′ if and only if a H a′ is true. On the other hand, if H is an asymmetric relation, the lines have arrows and are called **directed arcs**. In this case, there exists a directed arc with orientation from a to a′ if and only if a H a′ is true. When both a H a′ and a′ H a are true there exist two directed arcs, one with orientation from a to a′ and the other with orientation from a′ to a.

Whether a graph is directed or not (i.e., whether it contains directed arcs or undirected edges), the location of the vertices and the geometric representation of the lines (see Figure 7.2.1) will be influenced by a desire to make the diagram easy to read. (For more details on graph theory, see, for example, Roy, 1969-1970; Berge, 1973; Christofides, 1975.)

Figure 7.2.1: Graphical representation of an outranking relation on a set A containing 5 actions (the absence of an arc between vertices a_2 and a_3 corresponds to not a_2 S a_3 and not a_3 S a_2)

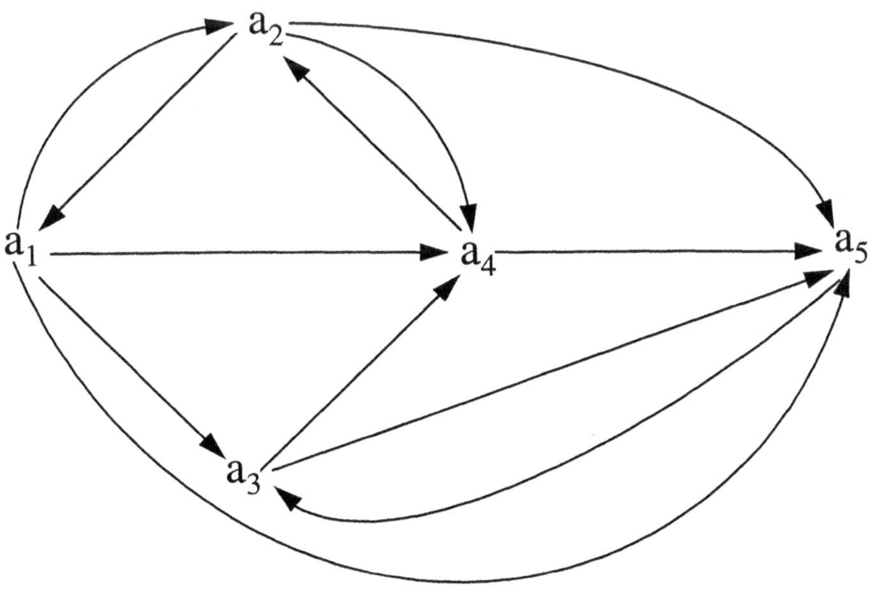

b) Notation for systems of preference relations

In general, a system of preference relations requires more than one binary relation. If only two conflicting relations are needed, and if at least one of the two is symmetric – e.g. a PSPR ($\succ$, $\sim$) or a BSPR (S, R) – a unique graph can synthesize the information contained in the relations. Let (H, T) be a system of two such relations, with T being symmetric. Since H and T are conflicting, a graph representing H alone would summarize the system; when the symmetric relation T holds between two vertices, H does not, and this could be shown by the absence of arcs between the vertices. Therefore, the graph of relation S is enough to represent a BSPR (S, R) (see Figure 7.2.1).

Summarizing the information of a system comprised of two relations that either do not conflict or are both not symmetric will normally require graphing more than one relation. This will also be the case when dealing with more than two relations – e.g., BSPR (I, P, Q, R) or CSPR (I, S, R). The information in these systems can be easily represented, however, by differentiating the lines (undirected edges or directed arcs) connecting two vertices according to the various binary relations. This convention complicates the graphical representation, but only slightly. We shall continue to call such diagrams graphs and use the notation defined in Figure 7.2.2.

Figure 7.2.2: Graphical conventions

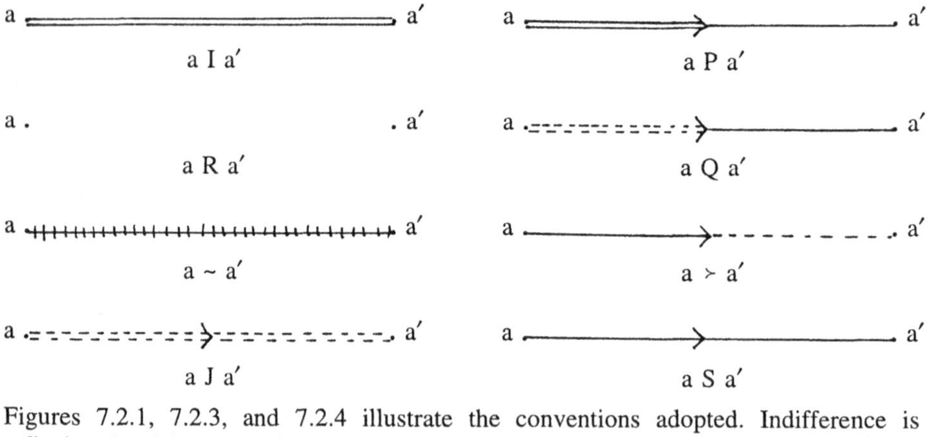

Figures 7.2.1, 7.2.3, and 7.2.4 illustrate the conventions adopted. Indifference is reflexive, implying a I a, $\forall$ a $\in$ A. To be rigorous, therefore, we should show an arc connecting each vertex to itself. Except where required for clarity or emphasis, however, we shall avoid showing these **loops**.

Figure 7.2.3: Representation of the system of preference relations in the family car example defined by (r 7.1.1) when opting for $a_n \; P \; a_f$

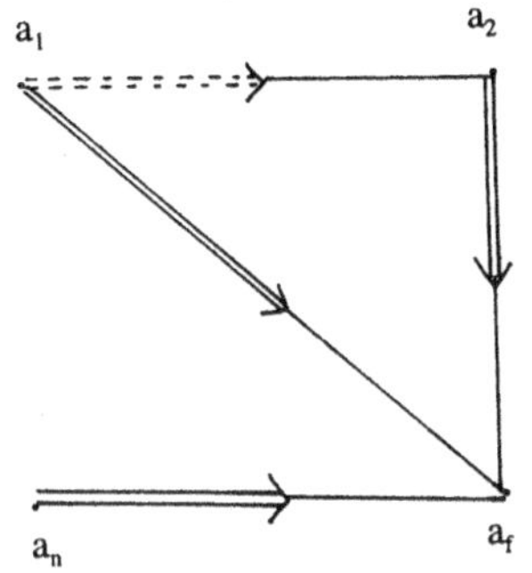

Figure 7.2.4: Representation of the basic system of preference relations (r 7.1.2) in the highway location example

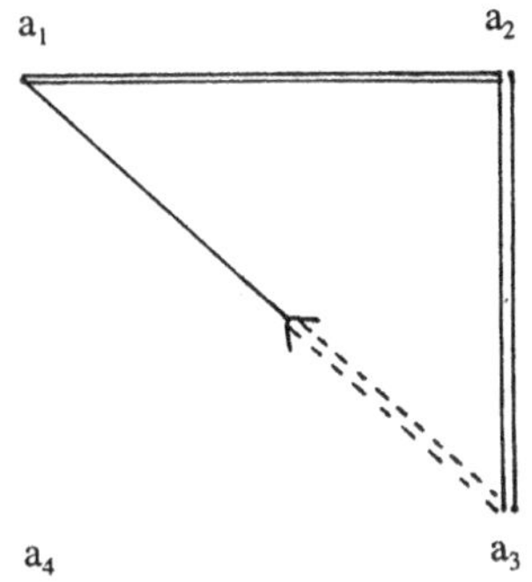

7.2.1.2 A new example: The mayor's preferences

At its next meeting, the municipal council of a small city V must discuss the pros and cons of four competing projects and support one of them. Unemployment is the chief concern of the council, as it has been estimated that between 11 and 12 percent of the 1500-1700 person potential work force is seeking employment. The four projects are all designed to address this concern.

Even though the municipal budget is tight, the council is ready to agree on financial assistance for projects that will create jobs for the unemployed of the community.

The four projects considered are such that the mayor Z, as well as the other members of the council, can compare any two of them based on two main aspects:

- Aspect No. 1: number of jobs created by the project (the type of jobs are similar for each of the projects);

– Aspect No. 2: cost of the project to the municipality (all the expenditures are to be incorporated in the next budget, which is to be discussed in the near future).

Evaluating the projects according to the second aspect is fairly easy. As it now stands, the first two projects would have identical costs for the municipality; the last two would also have identical costs, but between two and three times those of the first two projects (see Table 7.2.1).

Table 7.2.1: Possible evaluations for the council members
(p = probability of an unlikely event – on the order of one chance in ten;
c = approximately 10 % of the municipality's annual resources)

Aspects \ Projects	a_1	a_2	a_3	a_4
Aspect No. 1				
Number of jobs created	50 or 10	110 or 10	50	110 or 10
Corresponding probability	(p) (1 – p)	(p/2) (1 – p/2)	(1)	(1/2) (1/2)
Aspect No. 2				
Cost for the municipality	c	c	2 to 3 times c	2 to 3 times c

Evaluating the projects according to the first aspect is not as easy, however. The number of jobs that would be created by each project would largely depend on exogenous events that the council could not influence. The number of jobs could not, therefore, be predicted with certainty. Mayor Z has considered the different exogenous events and assigned the impacts and probabilities found in Table 7.2.1. All the actors believe these to be realistic estimates.

With only this information, many practitioners, researchers, and instructors in several European countries have played the role of Mayor Z in responding to questions designed to indicate preferences for the various projects (see Vincke, 1982).

Figure 7.2.5 illustrates the diversity of preference judgments obtained. The individuals involved all belonged to a European group involved with multi-criteria decision aiding and were all familiar with this type of experiment. We have also conducted the same experiment with other populations and obtained similar results. No matter how one tries to consolidate the information obtained in a CSPR, there remains a large disparity among individuals.

Figure 7.2.5: Enumeration of 30 responses to a survey concerning BSPR's

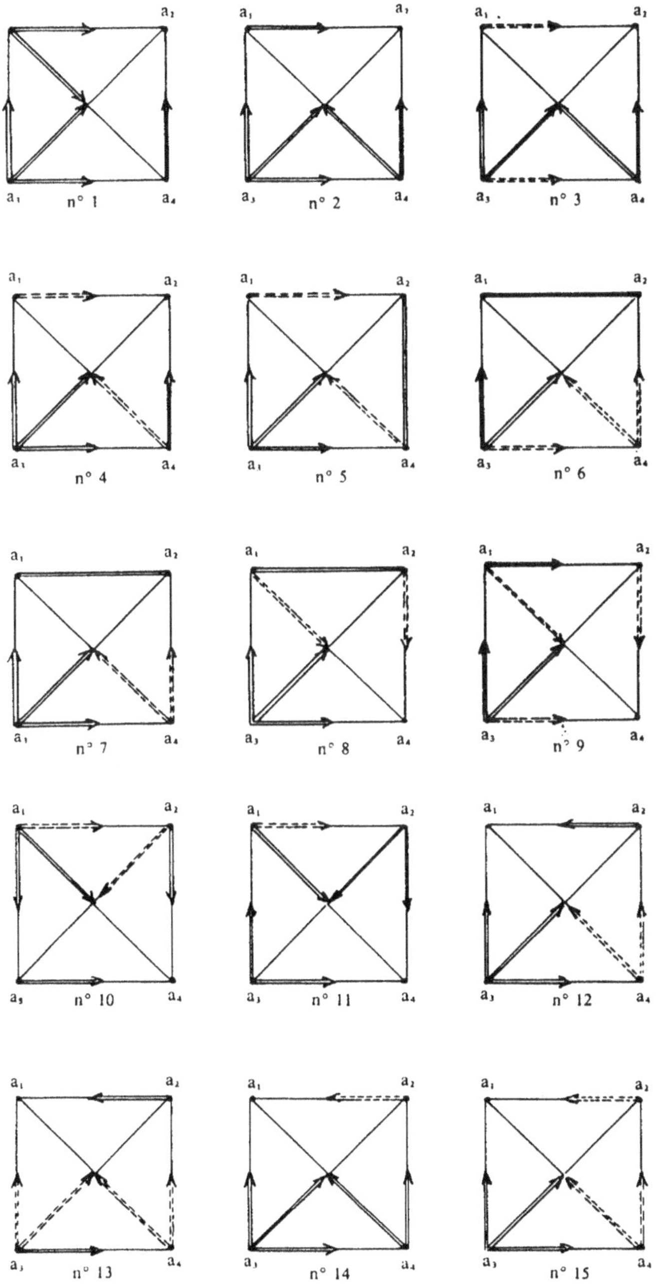

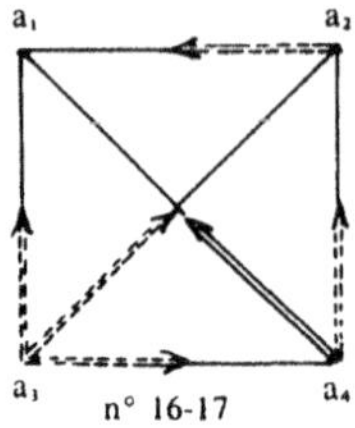

n° 16-17

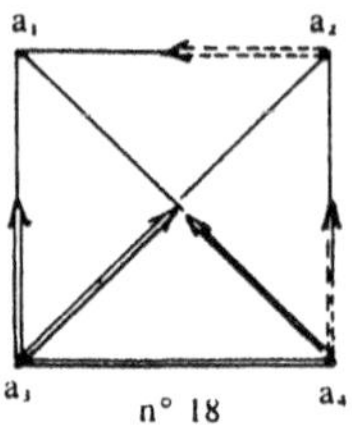

n° 18

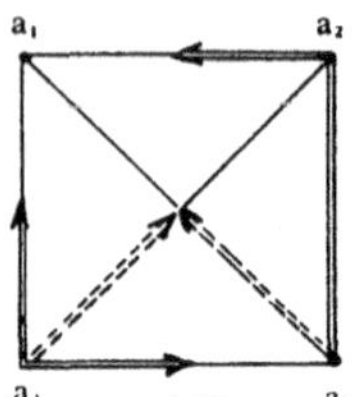

n° 19

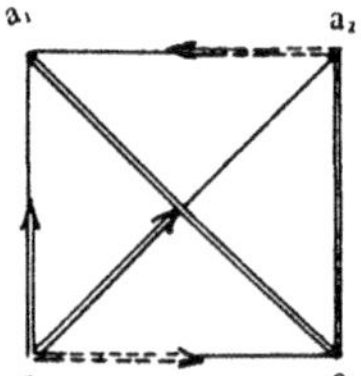

n° 20

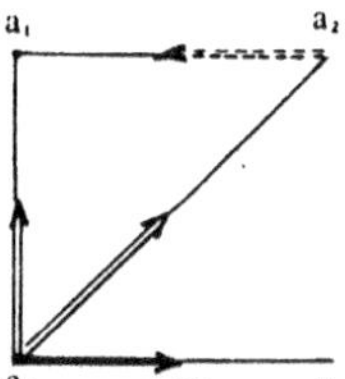

n° 21

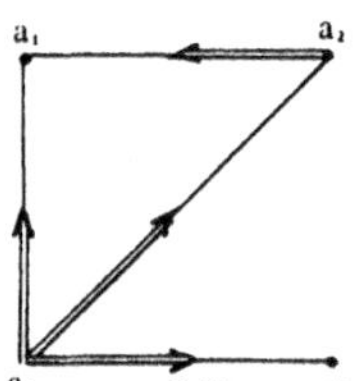

n° 22

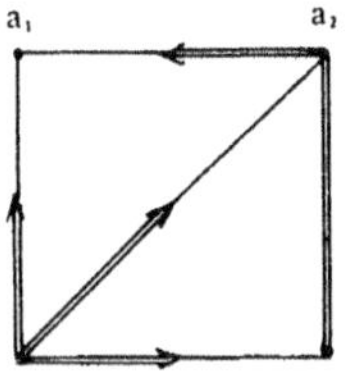

n° 23

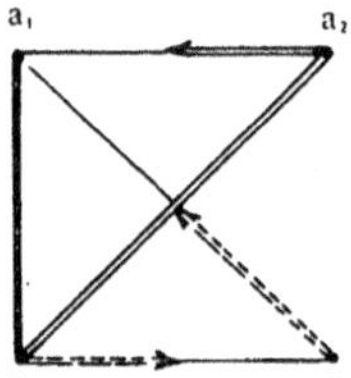

n° 24

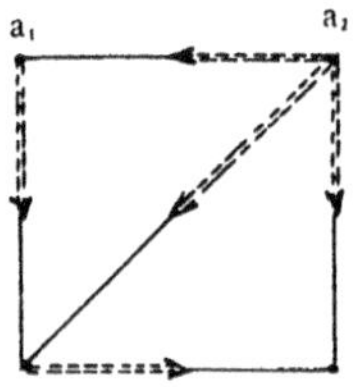

n° 25

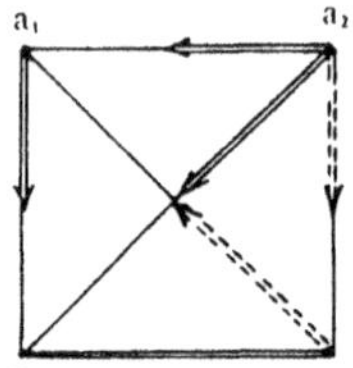

n° 26

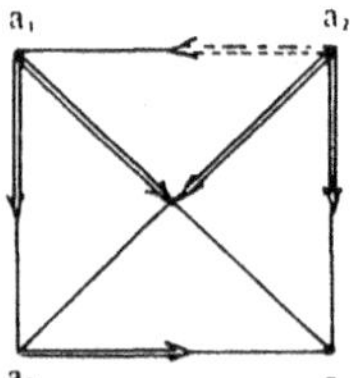

n° 27

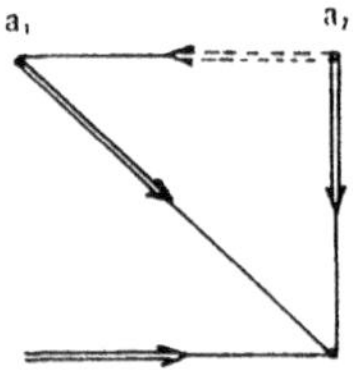

n° 28

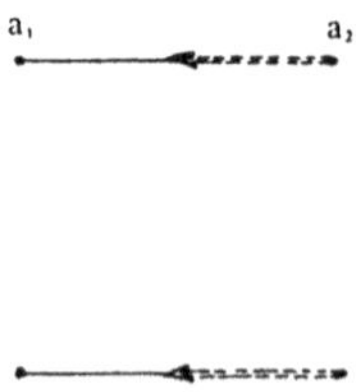

n° 29

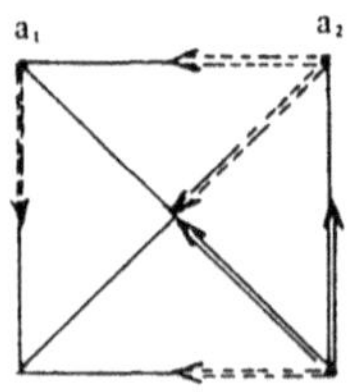

n° 30

7.2.2 Basic structures of SPR's that exclude or obscure incomparability

Although empirical SPR's will not always possess well-defined properties (see Figure 7.2.5), it is still useful to review the structures associated with SPR's of the most practical and theoretical interest. In this section we discuss the case where R = ∅, leaving the case where R ≠ ∅ for the next section. We first look at systems with one, then two, and then more than two relations. We shall not discuss relations that hold only for identical pairs of actions, as was the case with I in Section 7.2.2.1b), since these are of no interest here.

7.2.2.1 SPR's with only one relation

Technically, a single relation does not form a system. Nevertheless, let us consider single relations here. The unique relation can be symmetric, asymmetric, or neither. We now investigate briefly the structures associated with these three possibilities.

a) Equivalence classes

If the relation is symmetric, it must be either I or ~. Whichever is the case, the relation holds for every pair of actions in A,[11] and since there is only one relation, the case is trivial. It is clear that the relation is transitive and that all actions of A must be equivalent. The corresponding structure is an equivalence class (see Fig. 7.2.6; also Section 6.1.3).

Figure 7.2.6: Equivalence class structure: example of an SPR of the form (~)

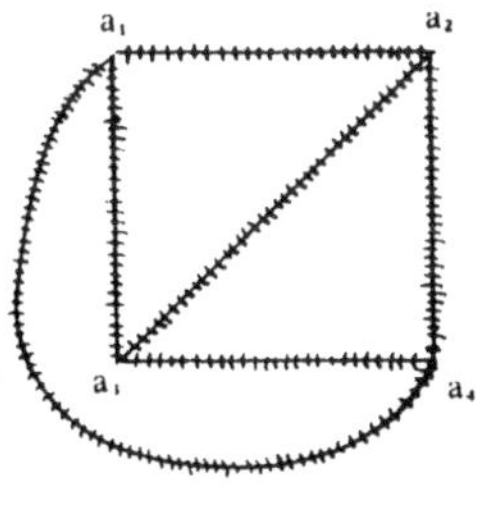

[11] *Recall that for any two actions a and a′, and any SPR (BSPR or CSPR), there exists a relation H of the SPR such that a H a′ or a′ H a.*

b) Complete orders[12] and intransitive tournaments

b1) Definitions

If the unique relation of the SPR is required to be asymmetric[13] (a relation frequently called a tournament), it must be either ≻, P, or Q, where an SPR with only Q is of little practical interest. In any case, we can consider only two different basic structures, one transitive and one intransitive. These two structures can be characterized by the absence or presence, respectively, of what are often called three-arc cycles.[14]

By definition, three actions a, a′, a″ form a **three-arc cycle** with respect to a relation V when the following three statements hold:

$$a \; V \; a'; \; a' \; V \; a'', \; a'' \; V \; a.$$

In an asymmetric relation, a three-arc cycle is incompatible with the transitivity of the relation; thus, these cycles can be thought of as special cases of **intransitive triangles.** They are only special cases, since other forms of intransitive triangles can exist in a relation that is not asymmetric; moreover, some three-arc cycles are compatible with transitivity (see Section 7.2.3).

On the other hand, the absence of three-arc cycles implies the transitivity of V. Therefore, we present the following two structures:

- **Complete order**: characterized by the absence of three-arc cycles (see Figure 7.2.8);
- **Intransitive tournament**: characterized by the presence of three-arc cycles (see Figure 7.2.7).

Figure 7.2.7: Intransitive tournament structure:
example of an SPR of the form (≻) with intransitive triangles

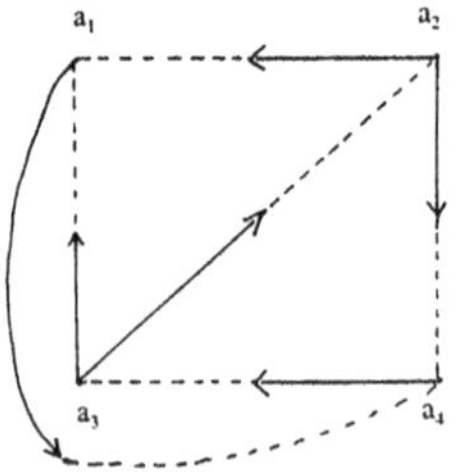

[12] *Complete orders are sometimes called strict orders.*

[13] *Rigorously, antisymmetric is not the same as asymmetric. To say that $\vec{H}$ is antisymmetric means that [a $\vec{H}$ a′ and a′ $\vec{H}$ a] can occur only if a = a′, while asymmetry does not even allow this exception. Given the rather artificial nature of hypotheses of reflexivity, we shall usually not need to worry about this distinction and shall use the two terms interchangeably.*

[14] *We note their importance in the works of Condorcet, 1785.*

Figure 7.2.8: Complete order structure:
example of an SPR of the form (P) without intransitive triangles

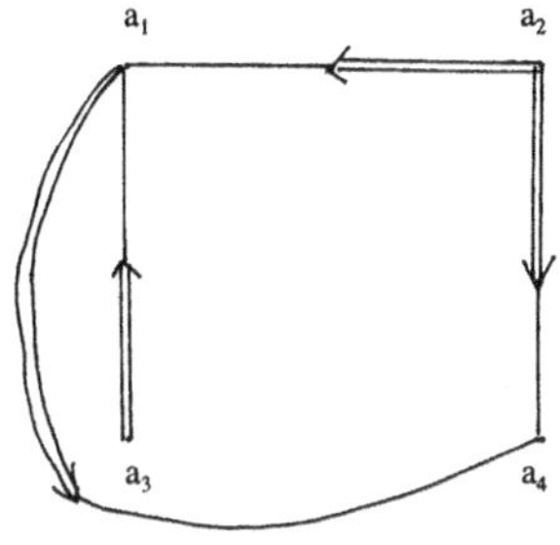

Example of a functional representation of P: $g(a_3) = 10$, $g(a_2) = 4$
$g(a_1) = 1$, $g(a_4) = 0$
Ranking function: $r(a_4) = 1$, $r(a_1) = 2$, $r(a_2) = 3$, $r(a_3) = 4$

b2) Functional representation of a complete order

The following is a simple but important result concerning the representation of an SPR by a function.

RESULT 7.2.1: For realistic problems,[15] *an SPR of the form V with a complete order structure can always be represented by a real-valued function g on A such that:*
$a' \; V \; a \Leftrightarrow g(a') > g(a)$. (r 7.2.1 a)

The function g is not unique. To see this, notice that when A is comprised of a set of m finite actions, the actions can always be arranged in an order $a_1, a_2, ..., a_m$ such that:

$$a_k \; V \; a_j \Leftrightarrow k > j.$$

Assigning numbers 1, 2, ..., m to the actions $a_1, a_2, ..., a_m$, respectively, defines a functional representation of the SPR V satisfying (r 7.2.1); but any other set of increasing numbers can be used. We shall call such a function g a **ranking function** (see Fig. 7.2.8).

No functional relation satisfying (r 7.2.1 a) can be found for an SPR V that has a structure of an intransitive tournament, since the values of g for three actions forming an intransitive triangle would have to satisfy:

$$g(a) > g(a') > g(a'') > g(a),$$

[15] *To be rigorous, we note that there are exceptions, but these are pathological situations that occur only when A is infinite, which is never the case in real problems. The reader can find examples of such situations and the necessary and sufficient condition for V to have the representation of (r 7.2.1) in Fishburn, 1970, pp. 26-29, for example.*

which is impossible.

c) Two-relation structures: a first look at complete basic systems of outranking relations (BSOR)

Finally, consider the case where the unique relation is required to be neither symmetric nor asymmetric, as is the case with S, J, K. Since the cases of J or K are of little practical interest, we shall only consider SPR's of the form S, which we have named complete basic systems of outranking relations (see Def. 7.1.5).

Consider first any binary relation H, which does not have to be complete. One can decompose this into two parts:

- a **symmetric part** $\mathring{H}$ defined by: a $\mathring{H}$ a′ ⇔ a H a′ and a′ H a;
- an **antisymmetric part** $\vec{H}$ defined by: a $\vec{H}$ a′ ⇔ a H a′ and not a′ H a.

As an example, the outranking relation S defined in Figure 7.2.1 can be divided into its symmetric part $\mathring{S}$ (reflecting situations of indifference), which holds for the three pairs of actions (a_1, a_2), (a_2, a_4), and (a_3, a_5), and its antisymmetric part $\vec{S}$ which holds for the couples (a_1, a_3), (a_1, a_4), (a_1, a_5), (a_2, a_5), (a_3, a_4), (a_4, a_5). (To determine S from $\mathring{S}$ and $\vec{S}$ presents no problem, even when S is not complete, as is the case here.)

So under these conditions, every complete BSOR (S) can be considered an SPR consisting of two relations – $\vec{S}$ (often easier to write simply as S), corresponding to the antisymmetric part of the basic relation, and $\mathring{S}$, corresponding to the symmetric part, which is I. Therefore, from a structural perspective, this single outranking relation is the same as an SPR with two relations – (I, $\vec{S}$) or (I, S) – where the first relation is symmetric, and the second is asymmetric. We now consider the principal structures corresponding to this type of SPR and, therefore, complete BSOR's.

7.2.2.2 SPR's with two relations

a) Complete preorders[16]

a1) Nonfunctional representation

Many readers will probably be familiar with a complete preorder, which was introduced in Section 6.1.3. It is still useful, however, to recall its most common forms.

---

[16] *Complete preorders are sometimes called weak orders, even though this expression applies only to the asymmetric relation.*

Consider a finite or countable family A_1, A_2, ... of nonempty subsets that are mutually exclusive and collectively exhaustive of A. Such a family is called a **partition of A**. Each of the subsets can be considered an equivalence class.

The easiest and most concrete way to characterize the structure of the complete preorder is to define such a partition and to rank the classes according to a complete order represented, for example, by increasing index values or by the order of proceeding from left to right along a line. To associate an SPR to any complete preorder structure (defined by the indices of the classes of a partition of A), we need only to introduce a binary symmetric relation T and a binary asymmetric relation V defined by:

$a' \ T \ a \Leftrightarrow$ a′ and a belong to the same equivalence class;
$a' \ V \ a \Leftrightarrow$ the difference between the index of the class containing a′ and the index of the class containing a is strictly positive. (r 7.2.1b)

Let (T, V) be an SPR made up of a symmetric and transitive relation T and an asymmetric and transitive relation V. The SPR (T, V) is said to have a complete preorder structure. To obtain the preceding representation, notice (see Fig. 7.2.9) that the properties of the relation T induce a unique partition of A and that the properties of V induce a complete order on the classes of this partition. This defines an SPR, since T is symmetric and complementary of V in the SPR considered.

Once again, note that given an SPR that forms a complete preorder, we can always combine the two relations T and V into one transitive[17] relation, H, defined as:

$$a' \ H \ a \Leftrightarrow a' \ V \ a \text{ or } a' \ T \ a.$$

Moreover, no information is lost in using the single relation H instead of the two relations T and V, since (using the notation of Section 7.2.2.1c) $T = \overset{\circ}{H}$ and $V = \overrightarrow{H}$ Also, if H is an SPR, the SPR will form a complete preorder if and only if H is transitive, since the transitivity of H leads to the transitivity of $\overset{\circ}{H}$ and $\overrightarrow{H}$, and vice versa.

a2) Functional representation

Consider a real valued function g defined on A. It is well known that this function forms a complete preorder on A. Simply place two actions in the same equivalence class if and only if they lead to the same value of g, and order the equivalence classes by increasing, or even decreasing, values of g.

[17] *The proof uses the fact that $a' \ T \ a \Leftrightarrow$ not $a' \ V \ a$ and not $a \ V \ a'$. We shall see in Section 7.2.3.2 that the relation H is not necessarily transitive in a SPR of the form (T, V, R) where T is symmetric and transitive, and V is asymmetric and transitive.*

We might also ask the complementary question, i.e., whether there exists at least one real-valued function g that can represent any given preorder structure on A in a simple and natural way. The following result is a reformulation of Result 7.2.1 applied to preorders.

Figure 7.2.9: Three representations of the same complete preorder structure on $A = \{a_1, a_2, a_3, a_4, a_5, a_6\}$

partition and order

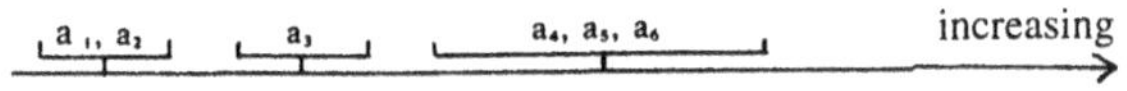

SPR (I, $\vec{S}$)

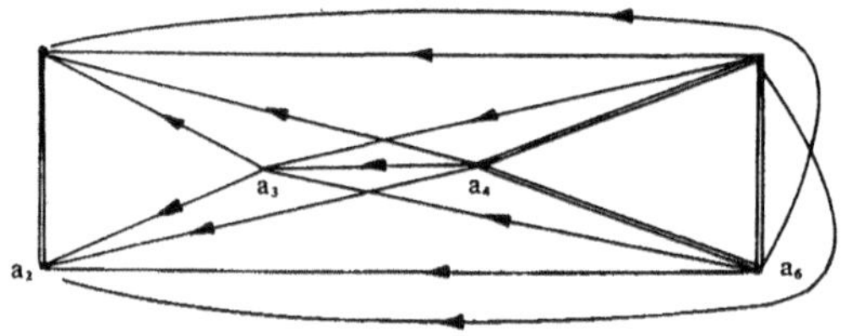

real-valued function g defined on A

$g(a_1) = 2$		$g(a_4) = 9$
	$g(a_3) = 2$	$g(a_5) = 9$
$g(a_2) = 2$		$g(a_6) = 9$

RESULT 7.2.2: For realistic problems,[18] *an SPR of the form (T, V) with a complete preorder structure can always be represented by a real-valued function g defined on A such that:*
$a' \; T \; a \Leftrightarrow g(a') = g(a);$
$a' \; V \; a \Leftrightarrow g(a') > g(a).$ (r 7.2.2)

One should keep in mind the arbitrary nature of the chosen representation. The example presented in Figure 7.2.9, for example, points up the arbitrary nature of the numbers 2, 5, and 9 that were chosen as values of the function g and highlights the fact that there are an infinite number of ways to represent this complete preorder structure by a function.

[18] *The cases where the result does not hold are the same pathological situations mentioned when presenting Result 7.2.1.*

b) Structure of a semi-order

The example of the cups of coffee with gradually increasing quantities of sugar presented in Section 7.1.2.4 was an example of a semi-order. This simple example demonstrated the nontransitivity of certain indifference relations that hinted at the importance of this rather little known structure.

b1) Example

Consider again the numerical data in Table 7.1.3 concerning the loan application. Assume that Z or someone acting in Z's name decomposes the set $A = \{b_1, b_2, c_1, c_2, a_1, a_2, a_3, a_4, a_5, a_6\}$ into the following six equivalence classes :

$$Y = \{a_4, a_5\}, X = \{a_6\}, C = \{c_1, c_2\}$$
$$N = \{a_2, a_3\}, B = \{b_1, b_2\}, M = \{a_1\},$$

and agrees upon the complete BSOR represented by the graph in Figure 7.2.10.

Although the BSOR in Figure 7.2.10 was based on a complete preorder defined on A, it cannot be identified with the complete preorder, since the indifference relation is not transitive. Making it transitive would lead to a poorer complete preorder comprising only two classes: one grouping Y, X, and C, and the other grouping N, B, and M. However, this could require Z to discriminate among small differences in performance levels of criteria when she does not wish to do so. Abandoning the transitivity of the symmetric relation in the definition of the complete preorder is what primarily accounts for the difference between this structure and that of a semi-order. Still, a certain amount of coherence between the two relations is required to define the semi-order. Unfortunately, the conditions expressing this coherence may be natural, but they are not straightforward.[19]

Figure 7.2.10: A complete BSOR having a semi-order structure on $A = \{b_1, b_2, c_1, c_2, a_1, a_2, a_3, a_4, a_5, a_6\}$ (see Table 7.1.3)

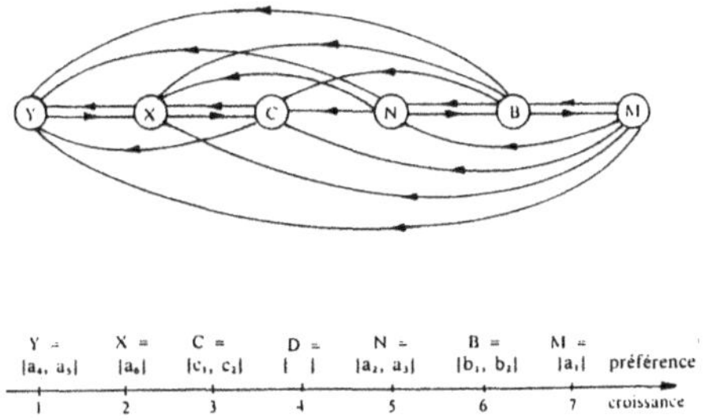

Two actions are indifferent if and only if they are placed in the same box or in two contiguous boxes.

---

[19] *translator's note: Page 140 in the original, French version discusses this further.*

b2) Semi-order properties

Consider a directed axis on which a number of boxes have been placed at the points corresponding to integer coordinates. Suppose that an actor Z has been asked to place each of the actions in a set A in one and only one of the boxes so that:

- she is indifferent between two actions if and only if fewer than q boxes separate the two actions;
- she prefers an action a′ to another action a if and only if a′ is in a box at least q boxes to the right of a.

The parameter q in this problem is called an **indifference threshold** (see Fig. 7.2.11).

The SPR (I, P) defined in this way possesses properties other than transitivity of P that are illustrated in Figures 7.2.12 and 7.2.13. Before formally stating these two properties, we need to recall the conventions traditionally used when combining relations.

To denote the existence of at least one action b such that c P b and b P a, we shall write c P^2 a. Similarly, c I^2 a will mean that there exists at least one action m such that c I m and m I a. For any two actions a and c, a semi-order does not allow the existence of two (distinct or not) actions (such as b and m above) such that c P^2 a and c I^2 a both hold. We write this (see Fig. 7.2.13) as:

$$P^2 \cap I^2 = \varnothing.$$

Similarly, consider two actions c and a, with c P a. To denote the existence of two indifferent actions b and b′ such that, c P b′ and b P a, we shall write c P I P a. The two prohibited relations shown in Figure 7.2.12 result, then, from a condition that we write as:

$$P\,I\,P \subset P.$$

Figure 7.2.11: Example of a semi-order structure with an indifference threshold q = 2

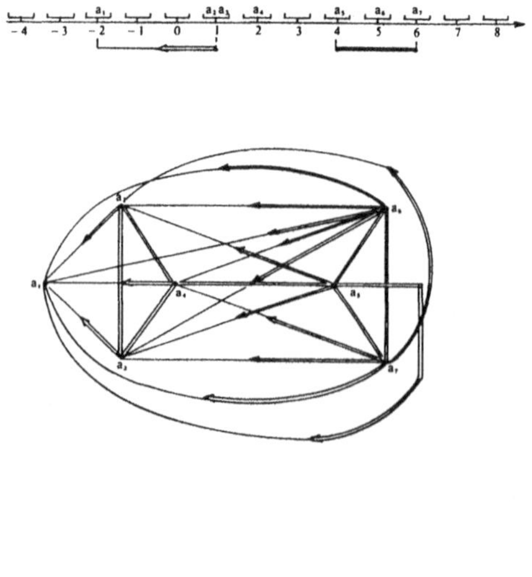

Figure 7.2.12: Illustration of the condition P I P ⊂ P satisified by all semi-orders

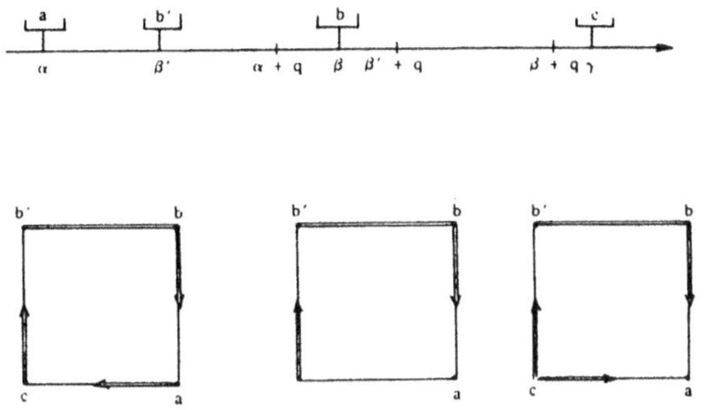

The two configurations are prohilited in any semi-order

c P b', b' I b, b P a
can only lead to this
configuration

Figure 7.2.13: Illustration of the condition $P^2 \cap I^2 = \varnothing$ satisfied by all semi-orders

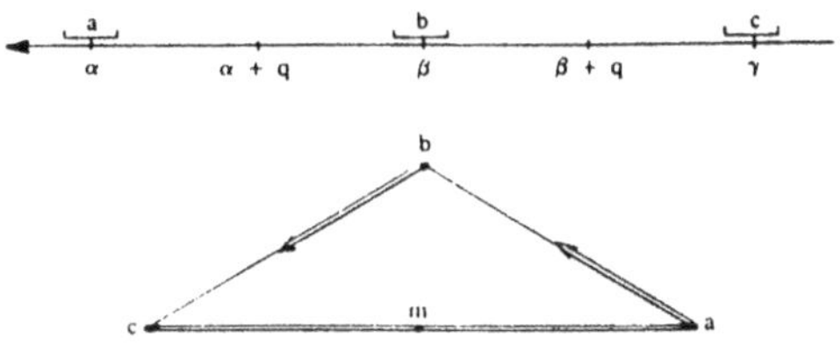

This configuration is prohibited in any semi-order

One might wonder whether these two conditions completely summarize I and P in ranking the actions of A in the presence of an indifference threshold q. It can also be shown[20] that, given a finite set A on which is defined an asymmetric relation P satisfying P I P ⊂ P and $P^2 \cap I^2 = \varnothing$ (where I indicates the symmetric relation defined by a I a′ ⇔ not a P a′ and not a′ P a), one can always place the elements of A in boxes along a linear axis and find an integer number $q \geq 0$, such that:

– a I a′ ⇔ there exist fewer than q boxes between those containing a′ and a;
– a P a′ ⇔ the box containing a′ is at least q boxes to the right of that containing a.

That is, P I P ⊂ P and $P^2 \cap I^2 = \varnothing$ are sufficient conditions for the existence of a semi-order.

---

[20] *For example, by directly establishing Result 7.2.3 below by using the proof of Result 7.2.7 in the French version, which is shorter than that of Fishburn, 1970.*

b3) Definition and functional representation

The preceding discussion and results motivate the following definition:

DEFINITION 7.2.1[21]*: An SPR possesses a* **semi-order** *structure if and only if it is of the form (T, V) with:*
i) T symmetric (and reflexive), V asymmetric;
ii) $V\,T\,V \subset V$;
iii) $V^2 \cap T^2 = \emptyset$.

Note the following:

a) The condition $V\ T\ V \subset V$ means that V must be transitive. Using the notation of Figure 7.2.12, it suffices to consider the example $b = b'$ to see that the condition implies $V^2 \subset V$.
b) A semi-order in which T is transitive is a complete preorder and, similarly, every complete preorder can be considered a semi-order in which the symmetric relation is transitive.

The following result formalizes in a slightly broader fashion the property presented at the end of b2).

RESULT 7.2.3: For realistic problems,[22] *an SPR of the form (T, V) with a semi-order structure can always be represented by a real-valued function g defined on A such that:*
$a'\ T\ a \Leftrightarrow -q \leq g(a') - g(a) \leq q;$
$a'\ V\ a \Leftrightarrow g(a') > g(a) + q,$ (r 7.2.3)
where q is a nonnegative constant[23] *called an indifference threshold.*

c) Other structures with one symmetric and one asymmetric relation

We present two additional structures that have one symmetric and one asymmetric relation. We define them after illustrating them through the use of special types of actions that will be used throughout the rest of this book.

---

[21] *For other equivalent definitions and for a deeper discussion of semi-orders, see Roubens and Vincke (1985).*

[22] *Again, the cases where the result does not hold are pathological cases not found in realistic problems. The reader can find details on this topic in Roubens and Vincke (1985).*

[23] *In Chapter 9, we shall present a more general representation using an indifference threshold q that is not necessarily constant. The properties that allow the use of a constant threshold instead of a variable threshold can be found in Roy and Vincke (1987).*

c1) Comparison of interval-actions

Consider a set of actions in which each action a_i is completely specified by two numbers, x_i and y_i, with $x_i \leq y_i$. We shall call such an action an **interval-action**. There are several ways to construct an SPR on a set of interval-actions A from the (x_i, y_i)'s. The following are two such ways.

First, define an SPR of the form (P, ~) by requiring:

$$(x_j, y_j)\ P\ (x_i, y_i) \Leftrightarrow x_j > y_i,$$
$$(x_j, y_j) \sim (x_i, y_i) \Leftrightarrow x_j \leq y_i \text{ and } y_j \geq x_i. \qquad \text{(r 7.2.4)}$$

That is, action a_j will be strictly preferred to action a_i only when the intervals corresponding to the two actions do not overlap and when a_j's interval is to the right of a_i's interval. It is clear that P is asymmetric and transitive in this case. The intervals (x_1, y_1), (x_2, y_2), (x_3, y_3), (x_4, y_4) shown in Figure 7.2.14 illustrate that $P^2 \cap \succ^2 = \varnothing$; therefore, (P, ~) does not form a semi-order.

Now define an SPR of the form ($\succ$, I) by requiring:

$$(x_j, y_j) \succ (x_i, y_i) \Leftrightarrow x_j > x_i \text{ and } y_j > y_i;$$
$$(x_i, y_i)\ I\ (x_i, y_i) \Leftrightarrow x_j \geq x_i \text{ and } y_j \leq y_i, \text{ or} \qquad \text{(r 7.2.5)}$$
$$x_j \leq x_i \text{ and } y_j \geq y_i.$$

Figure 7.2.14: Example of 4 interval-actions contradicting $P^2 \cap \sim^2 = \varnothing$ (see (r 7.2.4)) and $\succ^2 \cap I^2 = \varnothing$ (see (r 7.2.5))

Here, action a_j will be preferred to a_i when its interval is not completely included in that of a_i and is farther to the right. The relation $\succ$ is again asymmetric and transitive. And again, since the four intervals of Figure 7.2.14 contradict $\succ^2 \cap I^2 = \varnothing$, ($\succ$, I) does not form a semi-order either.

The structures of the SPR's presented as (r 7.2.4) and (r 7.2.5) differ in one way. In the first structure, one can show that $P \sim P \subset P$, while Figure 7.2.15 offers a counterexample to $\succ I \succ \subset \succ$ in the second structure. Therefore, in accordance with the following definitions, we shall say that ($\succ$, I) only forms a partial order, whereas (P, ~) forms an interval order.

Figure 7.2.15: Example of 4 interval-actions contradicting $\succ I \succ \subset \succ$ (see (r 7.2.5))

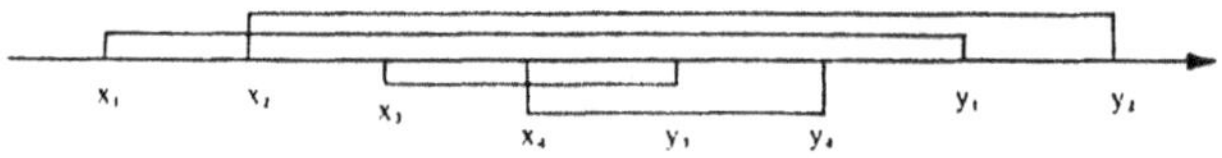

c2) Definitions and special cases

DEFINITION 7.2.2: An SPR of the form (T, V) with symmetric T and asymmetric V has:
- *a partial order structure if and only if V is transitive;*
- *an interval order structure if and only if $V\,T\,V \subset V$.*

A complete order is, therefore, a partial order in which T consists only of reflexive loops, and a semi-order is an interval order in which $V^2 \cap T^2 = \varnothing$.

7.2.2.3 SPR's with three or more relations

One can imagine a large number of systems containing three or more preference relations – (I, P, Q), (I, P, J), (I, $\succ$, J), (~, P, Q), (~, P, J), (~, $\succ$, J), (~, $\succ$, Q), (I, P, Q, J), (I, P, $\vec{S}$), ... We shall limit our discussion to those systems with three relations (call them T, V, W) where one (T) is symmetric, and two (V and W) are asymmetric, since other cases are of little general interest.

We begin with an example using interval actions but leave it to the reader to determine the structure required of the SPR in this case. We then introduce the pseudo-order, the structure of principal interest. Its importance will not become evident until Chapter 9, however. We finish by looking at a special case that is closely related to a semi-order structure, which we call a directed semi-order.

a) System (I, P, Q) on interval actions

Let us consider the difference between the two SPR's (~, P) and (I, $\succ$) defined by (r 7.2.4) and (r 7.2.5), respectively. In the first, any overlap of the intervals will imply indifference, whereas in the second, overlap only implies indifference if one interval is completely contained in the other. In many actual problems, the case where the two intervals overlap but where neither is included in the other can be treated as a case lying between strict preference and indifference. Therefore, the situation is one of weak preference. By adding Q to the SPR used previously, we have:

Figure 7.2.19: Example of an SPR of the form $(R, T, V) = (R, I, \vec{S})$ (the symmetric relation I and the asymmetric relation $\vec{S}$ are transitive, but $S = I \cup \vec{S}$ is not)

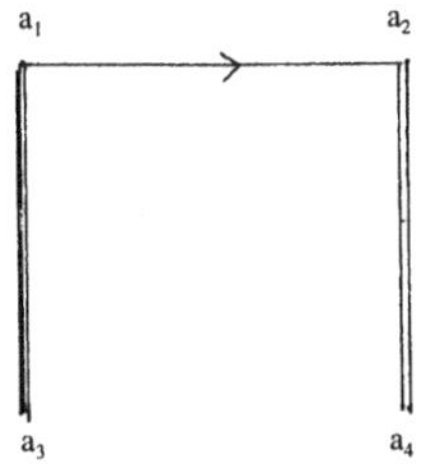

Note that in a partial preorder R is T-transitive but generally not transitive. The reader can easily see that:

- transitivity of $T \cup V$ implies transitivity of T and of V;
- if T is empty, (R, V) is a partial order when $R \neq \varnothing$ (see Def. 7.2.2, with R taking on the role of T), and a total or complete order when $R = \varnothing$ (see Section 7.2.2.1a);
- in this definition the condition, "$T \cup V$ is transitive," can be replaced by "T is transitive, V is transitive and T-transitive," or by, "T is transitive, V is transitive, R is T-transitive."

Let (R, T, V) form a partial preorder on A. Let us now consider whether or not this partial preorder can be represented by a real-valued function g on A where:

$$a' \; T \; a \Leftrightarrow g(a') = g(a);$$
$$a' \; V \; a \Rightarrow g(a') > g(a).$$

We cannot, of course, require $g(a') > g(a) \Rightarrow a' \; V \; a$, since this would imply $R = \varnothing$. To provide a valid representation of the partial preorder, then, we must also describe the conditions under which the two numbers g(a′) and g(a) ($g(a') > g(a)$) reflect a′ V a instead of a′ R a. We know of no simple solution to this problem, other than in special cases. That is, we know of no procedure that can use two numbers g(a) and g(a′) to determine whether or not a′ V a is true. Therefore, we know of no functional representation of the partial preorder.

It should be clear that if we assume that all pairs of incomparable actions are known, we could introduce a function, as we did when developing the functional representation of an order, whose sole purpose is to distinguish between the two relations T and V.

7.2.3.3 Other (R, T, V) structures

The most general basic systems of outranking relations exemplify the SPR's of interest here. These BSOR's can be written as $(R, \mathring{S}, , \vec{S})$. Therefore, we can denote $T \cup V$ by S in this subsection.

Figure 7.2.1 presents a good illustration of the type of SPR examined. The system represented there forms neither a partial order (since $\mathring{S}$ is not empty), nor a partial preorder (since $\mathring{S}$ is not transitive), nor a semi-order (since R is not empty). It is, therefore, a BSOR with none of the structures discussed previously. Yet, the system is realistic.

Two questions come to mind concerning an SPR of the form (R, T, V) with none of the properties discussed up to now:

1) to which structure is the SPR closest?
2) what noteworthy properties does it possess?[32]

7.2.4 Comparing preference differences or exchanges

We begin by illustrating the concept of preference differences through two examples which show that a new kind of action can be associated with any difference in preferences. These new kinds of actions can be considered exchanges of one action in A for another in A. The set of preference differences associated with these exchanges calls upon the concepts of indifference, preference, and incomparability.

a) Examples and discussion

Example 10: Application Packages (from Section 6.1.2)

Consider three application packages a, b, and c, and one evaluator Z who ranks them in the order a, b, c – i.e., a is better than b which is better than c. Z could also have an opinion on the quality of b relative to a and c. Assume that she feels that the quality of b is closer to that of c than to that of a.

This opinion is described by saying that the difference in the preferences separating a from b is greater than that separating b from c. These differences can be denoted, respectively, a $\ominus$ b and b $\ominus$ c, and we write:

$$(a \ominus b)\ P^{*}\ (b \ominus c),$$

where the relation P^{*} represents the subjective preference, "Z would rather go from b to a than from c to b."

Consider now a fourth candidate d, who has excellent test scores (which would make him or her slightly preferred to a, all else being equal), but poor grades (which would situate him or her on an equal plane with c). Z finds it difficult to compare d with the intermediate b, and until receiving more information, she states b R d. It is, therefore,

[32] *translator's note: The original, French version addresses these questions further on Pages 159-162.*

likely that it will be very difficult for Z to compare the differences $b \ominus d$ and $d \ominus b$ to the more favorable differences $b \ominus c$ or $a \ominus b$. More formally, we write:

$$(b \ominus d)\ R^*\ (b \ominus c);\ (d \ominus b)\ R^*\ (b \ominus c);$$
$$(b \ominus d)\ R^*\ (a \ominus b);\ (d \ominus b)\ R^*\ (a \ominus b).$$

Z can, nevertheless, consider the difference between a and c large enough to say $(a \ominus c)$ P^* $(b \ominus d)$ and $(a \ominus c)$ P^* $(d \ominus b)$.

(continued in Section 9.4.1)

Example 12: Plant organization (from Section 6.2.2)

Consider two assembly plants identical in everything but their production systems. The first plant has production system a_1; the second has production system a_2. Let b_1 and b_2 be two new production systems, considered to be improvements over a_1 and a_2, respectively. Assume that b_1 and b_2 are competing schemes. The question is whether an actor Z prefers to change Plant 1 from type a_1 to type b_1 (denoted, $a_1 \rightarrow b_1$) or to change Plant 2 from type a_2 to type b_2 (denoted, $a_2 \rightarrow b_2$). The symbolic statement:

$$(b_1 \ominus a_1)\ I^*\ (b_2 \ominus a_2)$$

indicates that Z is indifferent between the change (transformation, substitution) anticipated in Plant 1 ($a_1 \rightarrow b_1$) and that anticipated in Plant 2 ($a_2 \rightarrow b_2$). Another actor might feel that the change anticipated in Plant 2 is certainly not better than that anticipated in Plant 1 without feeling sure, however, that she prefers the change in the first. Such a judgment is formally written as:

$$(b_1 \ominus a_1)\ Q^*\ (b_2 \ominus a_2);$$

or

$$(a_1 \rightarrow b_1)\ Q^*\ (a_2 \rightarrow b_2).$$

(continued in Section 9.4.2)

These examples illustrate that in addition to Z's preferences for actions in a set A, there are many cases in which it is interesting, indeed necessary, to consider Z's preference differences between pairs of actions in the set. **The difference $a \ominus b$ can always be associated with an exchange, written $b \rightarrow a$, of action b for action a**. We shall continue to use one or the other of these equivalent notations, depending on whether we wish to emphasize the interpretation of preference difference or that of exchange of actions.

Even though the exchanges can be represented as actions (in $A \times A$), it is not always easy to conceive of such actions. It will, therefore, often be difficult to develop preference judgments for these exchanges or to interview Z about how she compares them. Instead of proceeding directly, as in the two previous examples, it is often useful to refer to an external dimension as is done in the following questions:

- does transforming action a into action b lead to a greater, equal, or lesser preference than transforming action c into action d?
- would you be willing to pay more, less, or the same amount to transform action a into action b than you would to transform action c into action d?
- if you could transform one action into another by performing a boring task, a task that offered no other reward than that of transforming actions, would you be willing to work more time, less time, or the same amount of time to transform action a into action b than you would to transform action c into action d?

b) Preference relations on A × A

As shown in the previous examples, one can use the basic relation of indifference, strict preference, weak preference, and incomparability to compare actions on A × A. We denote these I^*, P^*, Q^*, and R^*, respectively, the asterisk indicating that the comparisons are between preference differences or exchanges of actions rather than between actions of the original set A. Using the same convention as that for relations pertaining to A, we shall use the consolidated relations (see Table 7.1.5) $\sim^*$, $\succ^*$, J^*, K^*, and S^*.

Let H^* be any of the nine relations considered above. First of all, we note that, except in very special cases, H^* cannot be deduced from H. The only properties of H^* that follow from our discussion of relational properties on A are those of symmetry, asymmetry, reflexivity, and irreflexivity; and the interpretations of these properties for H^* are straightforward. The structure of the Cartesian product in A × A gives rise to properties of H^* that have no equivalents when considering H, however. We present three of these properties here and leave it to the reader to decide whether their interpretation for H is trivial or overly restrictive.

$$(a \ominus b)\ H^*\ (c \ominus d) \Rightarrow (d \ominus c)\ H^*\ (b \ominus a); \qquad \text{(r 7.2.9)}$$
$$(a \ominus b)\ H^*\ (a \ominus a) \Rightarrow (a \ominus b)\ H^*\ (c \ominus c); \qquad \text{(r 7.2.10)}$$
$$(a \ominus b)\ H^*\ (c \ominus d) \Rightarrow (a \ominus c)\ H^*\ (b \ominus d). \qquad \text{(r 7.2.11)}$$ [33]

[33] *translator's note: The original French version discusses some connections between relations on A and those an A × A and structural properties associated with preferences on A × A on Pages 165-168. See also Vansnick (1990).*

Chapter 8

COMPARING ACTIONS AND MODELING CONSEQUENCES

SUMMARY

Constructing any of the systems of preference relations on A requires a model of the information that affects the formation, justification, and evolution of an actor's preferences. This information is rarely available in a well-structured, quantified, or organized form, and what the analyst can use is often subject to imprecision, uncertainty, and inaccurate determination. In this chapter, we propose a methodology for approaching this phase of the modeling effort.

In Section 8.1.1, we define the term "consequence of an action" (Def. 8.1.1) to denote the various elements (effects, attributes, aspects, ...) that can interact with the objectives or value system of an actor and affect how she builds, justifies, or transforms her preferences. Our methodology is designed to analyze and distinguish the various consequences according to their quantitative and qualitative influences on the comparison of actions. Before the modeling effort begins, these consequences are ill-defined and possess fuzzy boundaries. They stem from complex and highly interwoven entities. At this stage, we refer to the consequence cloud.

In Section 8.1.2, we show the breadth and general nature of the approach used to isolate and define what we call elementary consequences (Def. 8.1.2) and offer concrete examples and practical illustrations. An elementary consequence usually points out the existence of an underlying dimension that reflects a preference shared among the different actors. This leads to the two basic concepts of Section 8.1.3: a preference scale (Def. 8.1.3) and a preference dimension (Def. 8.1.4). We present various examples and illustrations of these definitions in Section 8.1.3.2.

For a dimension to be operational, one must be able to map the impacts of a potential action on this dimension into a state or group of states with the help of some procedure. The procedure could be an empirical rule, a mathematical formula, a survey technique, or an experiment. This idea is the subject of Section 8.1.4 and leads to the concept of a state indicator (Def. 8.1.5) and the distinction between point and nonpoint state indicators. We end Section 8.1.4 with a discussion of the set of dimensions, which we call the consequence spectrum (Def. 8.1.6), that is used to describe the consequence cloud. Section 8.1.5 illustrates this first aspect of the methodology concerned with evaluating actions in the continuation of Examples 3, 5, and 6.

In Section 8.2.1, we discuss the deficiencies of using only point state indicators. These deficiencies are related to a lack of knowledge about the consequences of actions. The concept of a dispersion index is introduced to help model complementary information that can help portray the imprecision, uncertainty, and inaccurate determination associated with the consequences.

We introduce and illustrate the concept of dispersion thresholds in Section 8.2.2. We explain the difference between a nonpoint state indicator and a point state indicator with a threshold and define positive and negative dispersion thresholds associated with a point state indicator. We finish Section 8.2.2 by discussing the important difference between intrinsic and nonintrinsic dispersion thresholds.

The dispersion indicator that represents thresholds is in fact a special case of a category of dispersion indicators that we call modulation indicators. In Section 8.2.3, we illustrate four types of modulation indicators and provide a general definition (Def. 8.2.1).

Section 8.2.4 is devoted to a more general form of dispersion indicator: the referenced dispersion indicator.

We end the chapter with Section 8.2.5, where we summarize the importance of the different components of the evaluation model Γ(A). We also summarize the principles that should guide determination of such a model for any problem.

8.1 CONSEQUENCES OF AN ACTION, DIMENSIONS, AND ASSOCIATED STATE INDICATORS

Except in the case of P.δ, decision aiding requires a certain amount of preference modeling. The preferences to be modeled can be those of all or some of the actors concerned with the decision process. Whatever the form of the modeled preferences – systems of preference relations or related forms presented in the previous chapter – the preference model is based not on the various actions themselves, but on the consequences that result from the actions and on the different actors' subjective evaluations of them. Unfortunately, the data upon which the preference model is built are rarely available in a form that the analyst can use directly, even when considering a single actor Z. If the data did somehow appear in the desired form, one should be suspicious of their integrity (see Section 8.2).

After discussing the form in which data are usually available, we present a methodology in Sections 8.1.2-8.1.4 to assist the analyst in analyzing and organizing the data, and, especially, in using the available informational system to select, structure, and specify the basic elements from which the various actors can build, justify, or transform their preferences. Finally, we use three reference examples to illustrate the methodology in Section 8.1.5.

8.1.1 The consequence cloud

To say that one action is better or worse than another, or to say that an action is good or bad according to some norms, only makes sense in referring to the consequences (also called attributes or outcomes) of the actions and to the subjective value judgments made of these consequences.

*DEFINITION 8.1.1: Any effect or attribute of action a that can interact with the objectives, strategy, or value system of an actor involved in the decision process by serving as a basic element that allows her to build or change her preferences is called a **consequence** of a.*

Before discussing this definition, we emphasize that an effect or attribute that one actor considers essential might be considered insignificant, or even irrelevant, by another. More generally, the manner in which a particular consequence combines with others to determine comprehensive preferences will not be the same for the different actors. These differences often stem from different value systems (see Section 4.1). To determine what all the actors initially agree on, the proposed methodology analyzes and distinguishes consequences as a function of their qualitative and quantitative influences when comparing actions (see Sections 4.2.2 and 4.2.3). This chapter proposes a means of formalizing the elements as objectively as possible through the use of measurements, benchmarks, observations, opinions, ... The aim of this objective formulation is to make the evaluation model as general and transferable as possible.

We cannot really speak of the consequences of a single action, *a priori*, as this would imply that there pre-exist unambiguous and perfectly separable consequences. It is not usually obvious whether a given effect should be considered pertinent or whether it is related to or separate from other effects. That is, the consequences are not well-defined at the outset; their boundaries are vague, and they interact with each other in complex ways. We, therefore, refer to the **consequence cloud** and shall denote this cloud as $\upsilon(a)$ for the action a. To illustrate this notion, consider again the three examples of Section 7.1.1 (see also Roy and Bouyssou, 1986; Bana e Costa and Dinis das Neves, 1989; Ostanello, 1990; McCord *et al.*, 1993; Roy and Bouyssou, 1993, chapters 8 and 10; Maystre *et al.*, 1994).

The Family Car Example

The anecdote presented at the end of Chapter 1 leads to a few well-defined consequences that are used to compare two models of automobile. For the father, these consequences are purchase price, operating cost, comfort, and safety. For the other family members, aesthetic aspects and seating capacity must also be incorporated in $\upsilon(a)$. But since the seating capacity of a car would be closely related to its comfort, it might not be necessary to consider seating capacity and comfort as two separate consequences.

Note that none of the preceding attributes seems capable of explaining the daughter's preference for model a_f. Perhaps her preference stems from the prestige associated with the model or from its acceleration capabilities. Under these conditions, the impression a car would make on the neighbors or its acceleration capabilities would have to be considered in $\upsilon(a_f)$, and therefore, in $\upsilon(a)$, whatever the model of a.

Highway Location Example

In this case, $\upsilon(a)$ is much more complex. The principal consequences would be:

- effects on traffic: e.g., impacts on flows and delays;
- effects on residents: e.g., expropriations of land, noise;
- other effects on the environment: e.g., deforestation, damage to historical or scenic sites;
- costs: e.g., investment costs, maintenance costs;
- demographic effects: e.g., urban sprawl, impact on zoning patterns;
- political impacts: e.g., formation of opposition groups, pressure from potential users.

Loan Application Example

Assume that any clerical problems associated with an application can be ignored, and suppose that the financial organization has ample funds to satisfy the demand. The consequences to consider are, thus, the attributes of the applicant that would indicate his or her ability to repay the loan under the expected conditions. The ratios discussed in Section 7.1.1 were conceived with these attributes in mind. More generally, any information (regardless of its origin) that would shed some light on the applicant's solvency can be used as an element of $\upsilon(a)$ according to the definition of consequence given above.

These examples, especially the last one, point up the fact that what we are calling consequences of a are not only the results of executing action a. We believe that there is a need to have a single expression for the various elements (effects, outcomes, attributes, ...) that interact with the objectives or value system of an actor by contributing to the construction, justification, or transformation of her preferences. We feel that the word consequences is best suited for this expression.

Rarely would the analyst possess a useful description of the consequences making up $\upsilon(a)$ at the beginning of a study; the available data would not usually be in a suitable form. A systematic investigation of $\upsilon(a)$, conducted with the assistance of the client, would be required. The investigation must be designed according to the type of information that would ultimately become available. Indeed, it is pointless to build a model that requires unavailable data.

At this stage, **the vagueness and complexity of the consequences** make the analyst's task especially difficult. **The consequences are often poorly defined or specified and present two sources of difficulty that the proposed methodology**[1] **attempts to overcome**: There are multiple consequences, and they are not known with certainty (see Roy, 1988). The rest of Section 8.1 addresses the problem of multiple consequences. Section 8.2 treats the analysis and handling of factors related to uncertainty, imprecision, and inaccurate determination.

8.1.2 Elementary consequences

8.1.2.1 General remarks on modeling the consequence cloud

The preceding examples demonstrate the diversity and complexity of the elements of $\upsilon(a)$. In light of the types of questions that could arise, the analyst must build a framework for incorporating the phenomena included in $\upsilon(a)$. **To the greatest extent possible, this model must**:

---

[1] *This methodology differs in several ways from that proposed by the American school, which is based on the concept of attribute and on utility theory (see, especially, Keeney and Raiffa, 1976; Bouyssou, 1984; Roy, 1990; Vincke, 1992; Roy and Vanderpooten, 1996).*

- be applicable to any action a ∈ A; i.e., the model must consider all the consequences of actions in A. We express this by υ(A), which emphasizes the fact that the model considers any effect or attribute of any action of interest (see Section 8.1.1, a_f in the family car example) and not only the categories of consequences common to all of the potential actions.
- cover all categories of consequences, even those that are specific to a single actor (e.g., the daughter in the family car example). This coverage is limited, of course, to the preferences of the relevant actors in the current phase of the study. We add here that the consequences considered by a given actor depend[2] on the information she has available, on her value system, and more generally, on her "representation" system.

Note that techniques such as Planning Programming Budgeting System (PPBS), Rationalisation des Choix Budgétaires (RCB), and other tree-based techniques have been developed to handle the complexity of υ(A) (see Lyden and Miller, 1967; Levy-Lambert and Guillaume, 1971; Cleland and King, 1983). In general, some systemic approach (Klir, 1972; Wymore, 1976; Le Moigne, 1977; Walliser, 1977; Chapman *et al.*, 1992) will often be useful to help the analyst get started, especially in difficult cases.

Whatever the approach taken and the techniques used, the study can lead to identifying and classifying a certain number of "elementary consequences."

DEFINITION 8.1.2: An ***elementary consequence*** *c is an effect or attribute that is recognized as a consequence with the two following properties:*

1. *It is described in such a way that the different actors understand why it could be important to at least one actor.*
2. *It is sufficiently understood to allow a precise description of its concrete manifestation after a potential action a is executed. This description may be in terms of the states associated with the consequence c.*

A vaguely or ambiguously defined consequence will lead to confusion among the different actors and will not allow a clear perception of the specific ways in which the consequence is manifested. Therefore, for the consequence to be part of the model υ(A), an additional investigation designed to satisfy the properties of an elementary consequence will be necessary. It may also be necessary to distinguish among several elementary consequences. In other words, to have an elementary consequence, one must specify precisely what is being considered and in what terms (states) it is being considered (see Section 8.1.5). Let E_c be the set of states associated with elementary consequence c. Executing a will generally, but not necessarily (see Sections 8.2.1d and 8.2.3.3), result in a single state of E_c.

[2] *In the sense of Crozier and Friedberg, 1977; Lesourne, 1977; Oury, 1983.*

8.1.2.2 Illustrations and practical considerations

To identify an exhaustive and nonredundant family of elementary consequences representing υ(A), the analyst can attempt to classify the consequences according to the level at which they arise. It is useful to distinguish among three levels:[3]

a) consequences at the individual level;
b) consequences at the group or institutional level;
c) consequences at the structural or functional level.

To illustrate Definition 8.1.2, consider the family car example (see Section 8.1.1). The impression that a given automobile model makes on the neighbors does not constitute an elementary consequence. The different actors' heterogeneous perceptions of how this effect would be manifested make it practically impossible to define the states correctly; this would, therefore, violate Property 2. Even if the states could be appropriately described, a given family member might be unable to compare them and, more generally, would have difficulty in placing them on a preference scale. Moreover, the father might very well state "e P e′" when comparing two states e and e′, while the daughter would state "e′ P e."

Section 8.1.3 shows that trying to capture this consequence in such a manner will lead to a dead end. If the impression on the neighbors must be considered in the model, one needs to analyze this consequence further. It could, for example, be decomposed into several elementary consequences, each one mapping into the different actors' preferences in the same way.

8.1.3 Scales and dimensions

8.1.3.1 Methodological perspective: Definitions and illustrations

With few exceptions (e.g., that just presented), the elementary consequences presented above point out the existence of an underlying dimension that reflects a preference that is shared by the different actors. This should not be surprising, since the effects or attributes in question are only considered when at least one actor thinks them important when comparing two actions.

Nevertheless, the connection between the factual elements characterizing two states of the same consequence and actor Z's judgment when considering them independently of all other consequences may be somewhat unclear, subtle, and contingent on the actor Z. Given the objectives of this stage of the methodology, it is important that this connection be as independent of Z as possible. Moreover, a clearer and more solidly established connection for each of the elementary components of υ(A) will lead to a better understood and accepted, and, therefore, more operational model. This motivates the next two definitions.

---

[3] *translator's note: The original, French version describes each level on Pages 178 and 179.*

DEFINITION 8.1.3: A ***preference scale*** *E is a set of states ranked according to a complete preorder,*[4] *denoted $\succeq$, with the following property. In comparing two ideal actions a and a', which corresponds to comparing two states e and e', respectively, of E ($\upsilon(a) = e$, $\upsilon(a') = e'$), any actor Z accepts:*

- *the situation of indifference a I a' when e and e' are equivalent in the preorder ($e = e'$);*
- *the situation of preference a $\succ$ a' when e comes after e' in the preorder ($e > e'$).*

A ***degree*** *of E is any equivalence class of the preorder; the scale can be thought of as the set of these degrees.*

A set of purchase prices of an object ordered by decreasing value is an example of a preference scale for any purchaser Z. Specifically, Z will prefer to purchase the object at price e (action a) rather than at price e' (action a') whenever e is less than e'. On the other hand, a set of shapes or colors of the object, even if clearly identified, would not generally form a preference scale. It would only be a scale if the purchasers valued the various shapes or colors in the same way, when these value judgments are considered independently of the other aspects of the object.

Consider now an attribute that has a physical measurement, such as length, power, or calorie content. A set of states that are defined and ranked according to such a measurement does not necessarily form a scale. As an example, consider a new detergent that a firm considers putting on the market, and let the physical measure be the quantity (e.g., height in a specific machine) of suds produced under given conditions. A consumer's preference will, in general, increase with the height of suds as long as the height is less than a certain amount. After the height becomes greater than this amount (which may be difficult to specify exactly), the preference will decrease with increased height. The ordering obtained from the various heights of suds will not always correspond to the preference ordering, and this physical quantity, therefore, does not define a preference scale.

Finally, consider the case where the state e is defined by a vector rather than a single number. For example, consider the vector of revenues and expenses over 20 years. Given a discount rate, the present value v(e) of the revenues and expenses forms a complete preorder on a set E of states e. (The states e and e' will belong to the same grade if and only if $v(e) = v(e')$.) If all the actors accept the preorder defined in this way, E is a scale. It is possible, however, that some actors will not accept the preorder, believing, for example, that a state consisting of vectors with net revenues close to zero each year is incomparable to states with the same present value but with large deficits in the early years and larger surpluses in later years. Similarly, the same preorder (defined by v) will not form a scale on E if there exists an actor with a strong or weak preference for a vector with net revenues that increase in time over a vector with the same present value but that has decreasing net revenues.

[4] *See Section 7.2.2.2a.*

Some consensus among the actors on the most basic terms of comparison is desirable. We look for this consensus in the expressed preferences among the states associated with an elementary consequence. The term **preference dimension** will be used for an elementary consequence such that the set of its states forms a complete preorder reflecting the preferences of the relevant actors. We shall also call this simply a dimension, except where this would be confused with other dimensions.

*DEFINITION 8.1.4: A **preference dimension** is an elementary consequence c such that the set E_c of states associated with it in the stage of the analysis considered leads to a complete preorder forming a scale.*

8.1.3.2 Types of scales and practical considerations

The following are typical scales that allow definitions of dimensions for various elementary consequences[5]:

a) Monetary scale M;
b) Discomfort scale P[6];
c) Complexity scale C;
d) Risk scale R;
e) Functional breakdown scale D[7].

The examples in Section 8.1.3.1 concerning the height of suds and the vectors of revenues and expenses illustrate the caution one must show when trying to form dimensions by assigning a set of states to elementary consequences. The family car example illustrates another frequently encountered difficulty that arises when developing a set of states for an elementary consequence in an attempt to form a dimension. In this example, consider that each actor wants a large enough car to transport his or her friends. But independent of any other consequence that may be correlated with the number of seats, the degree "6 seats" may be preferred to "9 seats," whereas "4 seats" may be preferred to "2 seats." As in the detergent example, the lack of correspondence between the orderings obtained by the physical scale and preference for the physical consequence makes the physical scale unsuitable as a preference dimension. Still, an underlying dimension can be uncovered in this type of case with the following general technique.

Let o be the ideal state describing the relevant objective (e.g., 6 seats, height of suds considered ideal). Suppose that it is possible to think of the relative proximity of each state to the state o: One might think of an implicit model or of a measurable distance,

---

[5] *translator's note: The original, French version describes each consequence on Pages 181-182. See also Vansnick (1990).*

[6] *translator's note: for "pénibilité," in French.*

[7] *translator's note: for "dysfonctionnement," in French.*

for example. The states could then be ordered by this proximity with the resulting order forming a scale and a dimension. In this way the highest degree of the scale "6 seats" might be followed by the degree that groups the states "5 seats" and "7 seats" if these two are considered equally "close to" o. The next degree may correspond only to the state "8 seats" if the individual considered this "closer to" o than all states except "6, 5, and 7 seats."

Another useful technique for overcoming the difficulties associated with the nature of the degrees of a dimension consists of building sub-scales and defining sub-dimensions. The states associated with the elementary consequence c are decomposed so that they can be recombined in degrees that can be appropriately ordered. Suppose, for example that c represents the inconvenience associated with making an urban walking trip, and that the state describes the trip. The length of the trip (in meters, for example) might not constitute a scale, since two trips of the same length may impose different inconveniences due, for example, to topographic features (hills, stairways, ...) and mechanical aids (elevators, moving walkways, ...). In addition to a sub-dimension "length," one could consider a second sub-dimension relative to the topography and its mechanical aids. For example, one could use the technique that considered proximity to an ideal state to construct a second dimension dealing with the topography and the mechanical aids. These two sub-dimensions could then be combined so as to define a comprehensive scale that would make c a dimension. Given the elementary level at which the sub-dimensions are combined and the fact that they are referring to the same dimension of inconvenience, there should then be few differences in value systems among the actors.

When the elementary dimension is already a multi-dimensional description (a vector), as in the case of the vector of revenues and expenses through time, the complexity of these techniques becomes prohibitive, and one would be hard pressed to define an appropriate dimension.

8.1.4 State indicators and consequence spectrum

Some consequences will be difficult to model and require careful attention to all of the concepts presented above. In many practical cases, however, the analyst can avoid precise definitions of the elementary consequences and their states and immediately introduce scales and dimensions. Indeed, the use of certain absolute scales (money, time, number of people, ...) or relative scales (percentage, probability, deviation from a norm, ...) is often obvious.

To be operational, an approach used to define dimensions must allow valid comparisons among different actions by comparing their impacts on the dimensions as represented by a state or a group of states. In the simplest case, all the actors agree that executing any action will result in a well-defined state of the elementary consequence considered. This assumes that the analyst can associate with each dimension some type of procedure – e.g., a mathematical formula, an empirical rule, a survey technique, an experimental procedure – that all the actors consider unbiased. This procedure must map each potential action into a subset of states along the dimension considered. This subset should be as small as possible and contain the state (or states) that in all likelihood will be obtained when the action is executed.

DEFINITION 8.1.5: Given a dimension c, the ***state indicator*** γ_c *is the procedure, rule, technique, or process that yields the degree (or degrees)* $\gamma_c(a)$ *containing the state (or perhaps the states*[8]*) in* E_c *which, in all likelihood, will be obtained if action a is executed. This state indicator is called a* ***point indicator*** *if* $\gamma_c(a)$ *is a single degree of* E_c *for any potential action a; otherwise, it is called a* ***nonpoint indicator****.*

It is possible that, for some action a, an actor Z does not accept the $\gamma_c(a)$ developed by the analyst. Assuming that the states E_c and state indicator γ_c are clearly defined and accepted by Z, the difficulty must lie in the basic data that the analyst and Z use to represent the action a.

DEFINITION 8.1.6: The set $\overline{\text{v}}$ *of the* $\overline{\text{n}}$ *dimensions that can operationally and exhaustively describe the consequences of* $\upsilon(A)$ *is called the* ***consequence spectrum****. This spectrum is relative to the stage of the analysis considered.*

The spectrum $\overline{\text{v}}$ is considered operational if either a state indicator γ_c or a dispersion indicator (see Section 8.2) can be defined for any c of $\overline{\text{v}}$.

One must be careful when developing an exhaustive set of dimensions. First of all, vague state indicators might make a class of phenomena seem much broader than is actually the case. As an example, when trying to estimate the decreased travel time compared to a reference variant in the highway location example, it is important to specify under what conditions this decreased time is calculated. Is it calculated at night, during 15-minute peak periods, under average conditions? If under average conditions, how are temporal variations – e.g., time-of-day, day-of-week, seasonal effects – accounted for? Will the state indicator assume that the traffic remains constant, or will it consider possible growth in traffic due to the new highway? Will the decreased travel time be calculated using one or several origin-destination pairs?

A second subtlety associated with the exhaustive nature of $\overline{\text{v}}$ concerns overlooking a dimension that might turn out to be important. The possibilities associated with automating the ticket distribution and fare collection tasks of a transportation system might be analyzed from technical and economic perspectives. Such an analysis would overlook certain aspects of $\upsilon(A)$ if the automation caused a feeling of insecurity in the passengers that, in turn, could lead to decreased demand for the service.

---

[8] *We mentioned the interest in this possibility at the end of Section 8.1.2.1 and shall come back to it in Section 8.2.1d.*

Finally, too much or too little data could cause the analyst to create redundant dimensions of $\upsilon(A)$ or to omit dimensions that are important. In Section 10.3, we return to the independence and various forms of dependence that can exist among certain dimensions of the spectrum.

We mention in passing that various statistical methods (see Benzecri, 1984; Bertier and Bouroche, 1975; de Montgolfier and Bertier, 1978), which are beyond the scope of this book, can be useful in choosing the effects or attributes to form υ or in determining the individual degrees when defining scales and dimensions.

8.1.5 Examples

Example 3: Agricultural Development (from Section 6.1.1)

Let $\underline{x}$ be a plan. The consequences to be considered in this example are the effects that implementing $\underline{x}$ can have in country $\wp$ on:

1) the balance of foreign trade;
2) the level of employment in the agricultural sector;
3) the independence in food supply.

The consequence cloud $\upsilon(\underline{x})$ "defined" in this way is obviously complex. It also includes aspects that are difficult to account for because of uncertainties and extremely diverse exogenous factors.

Given the macro-economic level of this stage of the study, the analyst might be satisfied with a model that provides a very gross description of $\upsilon(\underline{x})$, at least as a first approximation. Specifically, by making a few simplifying hypotheses, he can avoid addressing in detail factors of imprecision, uncertainty, and inaccurate determination. Indeed, such factors would not seem important in differentiating among the various plans at this stage.

The three effects listed above could, therefore, be considered elementary consequences. For this to be so, the analyst must specify the contents of each as precisely as possible (Def. 8.1.2, 1) so that their ranges and limits are unambiguous: which aspects are to be considered, which can be omitted, ... The analyst must also provide a description of how each of the three consequences would be concretely manifested if $\underline{x}$ were implemented (Def. 8.1.2, 2); i.e., he must describe the states that would represent the effects considered. Here we shall limit ourselves to defining the three elementary consequences as follows:

c_1: the foreign trade deficit in agricultural products, expressed in millions of dollars per year;
c_2: the number of persons employed in rural areas, expressed in thousands;
c_3: the degree of autonomy in food supply, represented by the percentage of calories consumed from agricultural products produced in country $\wp$.

The states of c_1 can be defined by integers. They form a discrete, completely ordered set that reflects preferences – the preference increases as the number (deficit) decreases. (This "inverse" relation could be avoided by defining the states in terms of a trade surplus and using negative integers.) Providing upper and lower bounds for possible values of such states would form a set E_1 that is finite and completely ordered. This set would form a (monetary) scale in which each degree is a unique state. The states of c_2 are of a similar nature, with only the meaning of the integers being different. Here, preference is increasing with increasing numbers. The analyst could use a scale E_2 in which the degrees correspond to tens of thousands of persons employed; assigning a state to a degree containing it could be defined, for example, by rounding the integer characterizing the state to the closest ten thousand. (It would be necessary to avoid ambiguity among integers ending in 5000's.) Finally, the states of c_3 are modeled by numbers between 0 and 1. These could be grouped into degrees, each degree being characterized by a whole number between 0 and 100 (indicating a percentage). Preferences increase with increasing percentages. Therefore, the set E_3 of integers between 0 and 100 is a scale.

Using Definition 8.1.4, it follows that c_1, c_2, and c_3 along with their respective scales, E_1, E_2, E_3, define three pertinent dimensions for the present stage of the study. For each of them, it is possible to define a function g_i:

$$g_i(x_1, \ldots, x_h, \ldots, x_n) = \gamma_i(\underline{x}),\ i \in \{1, 2, 3\},$$

that, $\forall\ \underline{x} \in A$, represents a procedure whereby one could calculate the degree of E_i expected to result if plan $\underline{x}$ is implemented.

For example, in the case of i = 1, the quantities produced when implementing $\underline{x}$ could be estimated by knowing the relation between crop yields and farming practices employed. From these quantities, imports and exports could be deduced, since the demand is considered known and fixed, and since it has been assumed that there will be no simultaneous importing and exporting of the same product (see Section 5.1.2). The trade deficit can then be determined by making hypotheses on the import and export prices of each product. The function g_1 could be modeled as being linear in the variables x_h. The coefficients of this linear function would only be considered approximations, of course, since the yields and, especially, the prices are subject to uncertainty.

The functions g_i form point state indicators (see Def. 8.1.5). For this stage of the analysis, we shall assume that $\upsilon(\underline{x})$ is correctly specified by using the three dimensions c_1, c_2, c_3 as the spectrum and by using the three functions g_1, g_2, g_3 as state indicators.

(continued in Section 8.2.3.2)

Example 5: Media-planning (from Section 6.1.3)

Let p be any advertising campaign plan. A large number of factors might initially be considered in comparing the merits of different periodicals chosen in plan p. Many factors are specific to the periodical alone – total readership of the periodical, characteristics of its readers, cost of an advertisement placed in it, its format, ... Some are linked to relationships that exist between the periodical and its readers – the readers' perception of the image of a, whether they are occasional or regular readers, ... Still others can only be ascertained in conjunction with certain characteristics of plan p – correspondence of the periodical with the target population of p (readers at whom the advertisement is specifically aimed), use of coupons in the magazine, ... These are the aspects, the attributes, the effects related to the couple (a, p) that form the consequences of any action a (a periodical) in A'_p (see Section 6.1.3). To give an idea of the multiplicity of these consequences and the complexity of $\upsilon(A'_p)$, we reproduce in Table 8.1.1 a list of consequences based on Abgueguen (1971). Let us look briefly at the elementary consequences, dimensions, and state indicators for a few of them.

Table 8.1.1: Consequences constituting $\upsilon(a)$ in Example 5, based on Abgueguen (1971)

Code	Effect – Aspect – Attribute
c_0	Reliability of the information or strength of the support
c_1	Total readership
c_2	Target correspondence
c_3	Cost per thousand targeted readers
c_4	Opportunities to be seen
c_5	Structure of support audience
c_6	Editorial context
c_7	Journal image for its readers
c_8	Advertising context (volume, layout, nature of advertisements)
c_9	Magazine format
c_{10}	Reproductive quality of prints
c_{11}	Support impact on distribution
c_{12}	Regularity of readership
c_{13}	Advertising penetration
c_{14}	Propensity for novelty
c_{15}	Journal image
c_{16}	Support attention value
c_{17}	Support penetration for certain forms of expression
c_{18}	Use of coupons
c_{19}	Support reader psychology
c_{20}	Diversity of services offered by the support
c_{21}	Support sales tendencies

The total readership of a periodical is the total number of people (men and women, young and old, rural and urban, blue and white collar workers, ...) who have looked at a copy of a given issue of the periodical, even if they only looked at it for an instant. Since an issue can pass among several people, the number $g_1(a)$ can only be estimated by means of surveys. Regardless of how $g_1(a)$ is obtained, c_1 would seem to be an elementary consequence in which the states are characterized by integers – thousands of people, for example. If two periodicals are identical in everything else, preference would increase with increased readership. Therefore, this is truly a dimension.

The individuals who make up the readership of a are traditionally divided, somewhat arbitrarily, into two populations. The first is that of those readers in the "target" population of p, which is usually defined by socioeconomic characteristics such as sex, age, income, lifestyle, ... The second population consists of all other readers, i.e., the nontargeted readers. The ratio,

"target readership/nontarget readership,"

is a good indicator of what can be called the target correspondence of the periodical. All other things equal, the advertiser would prefer a higher correspondence. A low correspondence might indicate that target readership is relatively low. Thus, c_2 is a dimension, and the target correspondence is the state indicator. It is generally difficult to evaluate the size of the target readership: It usually requires surveys and requires that the socioeconomic characteristics of the target population are clearly (even if somewhat arbitrarily) defined.

Consider now the ratio:

$$\gamma_0(a) = \text{target readership/target population.}$$

This ratio increases as the number of people who look at a and are targeted by p increases. This ratio represents the strength of the periodical in reaching the target population, and, we can consider γ_0 as the state indicator of this phenomenon.[9]

(continued in Section 8.2.2.1)

Example 6: Research Projects (from Section 6.1.3)

In this example, the analyst could use the three broad categories described below to describe the consequences of any research task a. For each of these categories, we suggest one or, perhaps, a few elementary consequences and scales that could lead to dimensions. We also provide examples of state indicators.

---

[9] *translator's note: Further details on possible consequences are presented at this point on Pages 189 and 190 of the original French version.*

a) Aspects related to necessary resources

The analyst must consider the resources that would have to be provided to perform the research task. These resources include personnel, equipment, energy, and money. Each of these could lead to several elementary consequences. The principal investigator of the research task is usually the one who specifies the required quantities of each of these resources.

At another level, the analyst must consider the probable financial costs of performing follow-on tasks in the same project, even though the action is defined as the research task currently considered. He would also have to consider future development costs that would be necessary to make research results useful to the organization *O*. The states that define these elementary consequences would usually be considered only to orders of magnitude:

- sums less than one million francs;
- sums on the order of millions of francs;
- sums on the order of tens of millions of francs;
- sums in excess of a hundred million francs.

The analyst might use a scale (rounded to the nearest integer, for example) to express the value of the ratio:

future costs/current costs,

where the current costs are the costs of the task calculated using standards based on the resources requested. This ratio would be a nonpoint state indicator, since the future costs may be estimated to be, for example, "between five and eight times the current costs." Assume that the format used in submitting requests (see Section 5.1.2) asks several questions about the paths to be followed in the event of a successful task. The answers provided to these questions could then be used to estimate the future costs.

If there exists an external funding source (e.g., the government or an industrial research sponsor) for some research tasks, a special dimension accounting for external funding might be considered. The external funding level would not be known with certainty when the funding decision on the research task is made, and the associated state indicator would be a non-point indicator.

b) Aspects related to anticipated results

One might first consider results that would be useful in penetrating new markets or in increasing shares of existing markets. This could result in specifying an elementary consequence as:

"size of additional markets acquired due to the success of the research."

As an example, the set of states that would characterize a dimension might be:

- increase in market size that would certainly justify much larger investments than those anticipated for the research and development;
- increase in market size justifying an investment three to five times that expected;
- an uncertain increase in market size that would depend on uncontrollable factors, but that would nevertheless justify the expected investments;
- increase in market size expected to be too small to recover the anticipated investments.

The effects related to anticipated results might also be considered at a more "strategic" level, rather than being strictly limited to a direct increase in market size. In addition to spin-offs in other markets, the research results might, for example, promote a positive image of the company, pave the way for certain joint ventures or developments, or promote some new strategic initiative. A second elementary consequence might, therefore, be related to:

"the role of the research results in supporting a strategic policy of *O*."

The states would again have to be qualitative, for example:

- significant positive impact on a strategic policy of the upper management of *O*;
- significant positive impact on a strategic policy of one of the divisions of *O*;
- no significant impact on any strategic policy of *O*;
- possible negative impact on a strategic policy of *O*.

For the two dimensions presented, it would be necessary to elicit the opinion of one or several experts to determine which state would best apply to a given research task. When the experts' opinions differ, and the difference cannot be explained by a difference in interpretation of the states, the state indicators would be non-point indicators.

c) Aspects reflecting uncertainty in results

The two elementary consequences presented in b) were developed as if the future research results could be anticipated. Experience has shown that organizations are uncertain about research results, even when the research is almost finished and close to entering the development stage. In general, it is unrealistic to believe that the different levels of results can be determined precisely enough to specify the outcome of a project. Even if this were feasible, it must also be possible to modulate the other consequences as a function of the final level obtained. For this reason it is more realistic to think of anticipated results characterizing (even if somewhat imprecisely) the "success of the project," and it would, therefore, be important to consider the plausibility of any assumptions made in forecasting this success. It would be difficult to identify the states of an elementary consequence:

"credibility of expected results, given the resources requested."

To do so would require a great deal of insight into aspects such as the intrinsic difficulty of the research, the capabilities of the research team, the adequacy of the available resources, and the research design. Our experience in several studies in this domain has led us to conclude that opinions on this matter cannot be validly quantified as probability distributions, given the time frame and environment of the real cases. Specifically, since by nature each research project is original, one cannot draw upon historical results, and the subjective opinions generated by equally respected experts can differ considerably. Therefore, a qualitative scale with only a few degrees remains the best solution in the majority of cases. As an example, the following might be considered:

- no reason to doubt the overall success;
- overall success of the project is doubtful, but some interesting results should be obtained;
- serious difficulties are envisioned that would lead to no useful results.

Using three degrees, as proposed in this scale, has a serious disadvantage, however: It can often lead to biasing opinions toward the middle degree. To reduce this phenomenon, one might develop finer degrees, for example:

- the risk of failure is considered extremely small;
- only unlikely circumstances or minor obstacles would cause actual results to differ from anticipated results;
- overall success of the project is doubtful, but some interesting results should be obtained;
- the risk of a total failure is small, but not negligible;
- serious problems make the final result highly unlikely.

We could eventually treat the first two states of this latter scale as if they belonged to the same degree and do the same for the last two states, and in this way recover the former three-degree scale. Using the latter scale initially, however, would reduce biases toward the central degree.

Other aspects relating to handling and predicting the consequences – the impact on future research, for example – could also be considered.[10]

(continued in Section 8.2.2.2)

[10] *translator's note: More detail is provided at this point on Pages 194 and 195 of the original French version.*

8.2 EVALUATING AN ACTION: DISPERSION INDICATORS TO MODEL IMPRECISION, UNCERTAINTY, AND INACCURATE DETERMINATION

In this section, we assume that the dimensions of the consequence spectrum are numbered from 1 to $\overline{n}$, and that $\gamma_i(a)$ is defined for all i, either as a single element of E_i (point state indicator) or a larger subset of E_i (nonpoint state indicator).

From the preceding section, one can see that in some cases $\overline{n}$ point state indicators may be sufficient to summarize the available information of $\upsilon(a)$ for every a in A, for all the actors considered, and for the current stage of the analysis. Thus, $\upsilon(a)$ can be modeled as an ordered series:

$$\gamma_1(a), \ldots, \gamma_{\overline{n}}(a) \in E_1 \times \ldots \times E_{\overline{n}}.$$

This forms what we call the "evaluation of a." This simplest form of evaluating an action is often inappropriate due to nonpoint state indicators or insufficient because more information is required. The **evaluation of a** then takes on a more complex form which, as we shall show, can always be denoted by:

$$\Gamma(a) = [\gamma_1(a), \delta_1(a)], \ldots, [\gamma_{\overline{n}}(a), \delta_{\overline{n}}(a)],$$

where $\gamma_i(a)$ is not necessarily a point indicator, and $\delta_i(a)$ is what we call the **dispersion indicator**, which represents the likelihood or some other type of modulation of the different states. This representation may be a probability distribution or a fuzzy membership indicator, for example.

Before considering the possible forms for these dispersion indicators in Sections 8.2.2-8.2.4, let us look more closely at why representing an evaluation by $\overline{n}$ point state indicators might not be satisfactory. The reasons are all related to a lack of knowledge about the consequences of a. The analyst might be able to avoid indicators of this lack of knowledge by making certain approximations or by cleverly defining the scales and state indicators (see Section 8.1.5). This may not always be possible, however, and it is necessary to have a formal procedure that can portray the imprecision, uncertainty, or inaccurate determination of the consequences.

In many cases, it is more than a simple ignorance that leads to this uncertainty, since ignorance would imply the possibility of eventually knowing, given enough time and sufficient resources. Only rarely, however, can the consequences of an action be measured in the same way that one can measure a mass or a time, or even in the same manner that other properties, such as temperature, can be relatively determined. The aspects of the reality to be portrayed are often impossible even to define objectively; attempting to do so seems to run into a kind of "arbitrary residual," which can only be handled by somewhat artificial conventions. The imprecision, uncertainty, and inaccurate determination, therefore, are fundamental, and the associated attributes exhibit what Watzlawick (1976) calls second order reality.

The general term dispersion indicator, then, represents a means to complete the information contained in the state indicator by making explicit the imprecision, uncertainty, or inaccurate determination of this information. These dispersion indicators can come in a variety of forms (e.g., thresholds, probability distributions, or fuzzy membership values), and we shall try to summarize by returning to the idea of evaluating an action at the end of this chapter.

8.2.1 Lack of knowledge and state indicator deficiencies

Whether the state indicator γ_i is a point or nonpoint indicator, it will often be insufficient to capture all the relevant information concerning the effect or attribute associated with the ith dimension. Let us briefly mention the situations that we have encountered most often.[11] Since these situations usually depend on the definition of the scale E_i, we shall assume that a scale has already been defined in each case.

a) The state indicator is coarser than the scale degrees

This situation is common when using a discrete scale with degrees that are very "close to each other" and is inevitable when using a continuous scale.

In many cases, the number $\gamma_i(a)$ can be calculated to as many digits as desired, creating the illusion that it is portraying a precise result. To fail to add some complementary information on the point indicator's degree of precision (e.g., plus or minus a certain number of degrees) would amount to omitting potentially valuable information for the decision aiding effort. Trying to handle this imprecision by combining different degrees of E_i so as to define a "less refined" scale will also lead to a loss of information in many cases. Indeed, even though one should avoid excessive refinement in developing the degrees of a scale (e.g., the purchase price to the nearest dollar), it is still usually better to use finer scales and add complementary information than to use coarser scales: Values on the coarser scale can always be determined from those on the finer scale, whereas the reverse is not true.

b) The degrees of the scale are defined imprecisely and overlap

Overlapping in the scale degrees is particularly noticeable in scales such as:

{very strong, strong, average, weak, very weak},
{A, B, C, D, F} (grades in an academic setting),
{very good, good, passable, fair, bad, very bad}.

The problems in assigning an action to a unique state stem not only from intrinsic difficulties in forming an opinion about the action, but also from the rather arbitrary and

[11] *translator's note: The original French version expands slightly on each of the situations on Pages 197-202 of the original French version.*

vague boundaries between very strong and strong, between C and B, between bad and fair, ...

It is better to use a nonpoint state indicator when the definition of the scale E_i makes such overlap possible. This would allow a depiction of a single evaluator's hesitations. When there is more than one evaluator, nonpoint indicators could be used to indicate the possible differences in opinion among the evaluators. Even so, using $\gamma_i(a)$ alone might not convey all the important information. Suppose, for example, that an individual evaluates action a to be "fair," but might also be convinced to accept an evaluation of "passable." Stating:

$$\gamma_i(a) = \{\text{fair, passable}\},$$

places the two evaluations on the same plane, contrary to what the individual really felt in supplying these judgments. As will be shown in f) below, such a loss of information becomes even more striking in the case of multiple evaluators.

c) A single state fluctuates in time or space

Consider the size of a well-defined physical quantity (e.g., the quantity of a certain substance in the blood of a patient), but which varies in time due to factors that are difficult to account for (foods ingested, anxiety, ...). A point state indicator based on the average or the median of the observations of the physical quantity would not reflect these fluctuations. A nonpoint indicator corresponding, for example, to the interval containing the extreme observations would also seem inadequate. So, there are situations where the imprecision is attributable neither to the state indicator itself (as in a)), nor to definition of the scale degrees (as in b)), but to the fact that the state of interest is not inherently stable.

d) The set of states is dispersed in time or space

To illustrate this situation, consider the noise level affecting a given population. The scale that is to be used can be either quantitative (e.g., decibel level or noise number index) or qualitative (see Alexandre and Barde, 1973). The noise level will reach a maximum in some area and decrease with the distance from this area. One might try to subdivide the overall study area into a number of zones. Each zone would contain an area where the noise level is fairly constant, and a zone could, therefore, be characterized by a degree of the scale. In general, however, the family of degrees defined in this way (nonpoint state indicators) would not be sufficient for evaluating the consequence of interest. The family would, of course, be divided into as many elementary consequences as there are zones, but this would assume that the determination of the zones would be independent of the action a to be evaluated. It would be simpler to supplement the different degrees belonging to $\gamma(a)$ with the numbers of inhabitants affected by the noise level corresponding to the degrees.

e) Some phenomena are not known with certainty, but can be probabilistically modeled

These are the situations corresponding to classic probability theory.

f) Differences in understanding or in opinions lead to doubt

When several point evaluations (e.g., each obtained from a different source) are assigned to a given scale E_i for the same action a, it may seem natural to consider the state indicator $\gamma_i(a)$ as a nonpoint indicator. Simply listing the states that resulted from the evaluations places all of the evaluations on the same plane, however, and could cause a loss in important information. Some states may have been proposed more often than others, or some evaluators may be recognized as having been better informed than others. Although this type of information is usually insufficient for assigning objective probabilities to the degrees of $\gamma_i(a)$, one could take advantage of this supplementary information to place the degrees in order of decreasing credibility. This situation, like those in d) and e), requires that the different degrees of $\gamma_i(a)$ be treated differently. Unlike d) and e), however, this modulation would usually have to be ordinal.

g) The state depends on the action of another state and is, thus, undetermined

This is the situation that the classical literature (see, for example, Ponssard, 1977) calls a "game." The primary behavioral features considered are usually categorized as different "strategies" of another well-defined individual. This individual need not be an actor considered in the analysis. The major difference between this situation and that corresponding to e) is the freedom available in developing the strategies (with or without intention to influence the decision) and the impossibility of determining objective probabilities for them by referring to statistical properties. Nevertheless, some of the strategies will almost always appear more likely than others. A ranking in terms of increasing credibility, perhaps even a subjective probability, for each of them, is usually worth considering. That is, even if the state of interest remains undetermined because of the potential actions of another individual, the information on the likelihood of the strategies of this individual could be useful in completing the definition of a nonpoint state indicator (see Section 8.2.4).

h) The state can only be determined by referring to diverse scenarios

Finally, there are situations in which the mutual action of numerous actors or a joint evolution of diverse factors affecting the decision environment can cause a great deal of inaccurate determination. Here again, the situation corresponds to one of a game, but the indetermination is no longer due to the will of some other well-defined individual. Again, one might wish to categorize the major possibilities, but they are now called "strategies of nature" or "scenarios" (see Section 6.1.3, Ex. 2).

8.2.2 Dispersion thresholds

8.2.2.1 Intrinsic and nonintrinsic dispersion thresholds

Consider a scale E_i where the degrees need not be numerical. Consider also a correctly defined point state indicator $\gamma_i(a)$. In situations like those described in Section 8.2.1a), b), or c), assume that an error analysis has allowed an estimation of the tolerance levels or of the fluctuations so that we can associate two additional degrees to each action a:

- $\gamma_i^-(a)$, which is a degree representing an underestimation of $\gamma_i(a)$;
- $\gamma_i^+(a)$, which is a degree representing an overestimation of $\gamma_i(a)$.

We can then define the two nonnegative quantities:

$$\eta_i^-(a) = \gamma_i(a) - \gamma_i^-(a),$$
$$\eta_i^+(a) = \gamma_i^+(a) - \gamma_i(a),$$

by subtracting the appropriate numbers if E_i is a numerical scale, or by counting the degrees of E_i required to climb either from $\gamma_i^-(a)$ to $\gamma_i(a)$ or from $\gamma_i(a)$ to $\gamma_i^+(a)$ if E_i is a nonnumerical scale.

We call $\eta_i^+(a)$ the **positive dispersion threshold** and $\eta_i^-(a)$ the **negative dispersion threshold**.

First of all, notice that the magnitude of these two thresholds may be completely unrelated. The fact that they are often equal (except at the ends of the scale, where one may be more "naturally constrained" than the other) should not lead one to automatically assume their equality.

Indeed, the relative likelihoods of certain errors or fluctuations will not necessarily be symmetric. (One can think, for example, of the asymmetry of actual costs for services about quoted costs when obtaining professional estimates.) Certain assumptions made in calculations or in sampling conditions may be considered either liberal or conservative. Certain "rounding off" options taken in defining $\gamma_i(a)$ may also lead to a lack of symmetry.

For example, consider a discrete scale E_i and a case where the state indicator is such that the imprecision will never produce a hesitation beyond two consecutive degrees. Assume that one wanted to be conservative and, therefore, constructed the state indicator so that it systematically gives the lower degree in the hesitations. Under these conditions, the negative threshold will always be equal to 0, whereas the positive threshold could either be 0 or 1.

Next, consider possible dependence between the thresholds and the action a. The simplest case is when this dependence appears only through the degree $\gamma_i(a)$. This means that given two actions a and a′ such that:

$$\gamma_i(a) = \gamma_i(a') = e \in E_i,$$

it follows that:

$$\eta_i^-(a) = \eta_i^-(a') = \eta_i^-(e);$$
$$\eta_i^+(a) = \eta_i^+(a') = \eta_i^+(e).$$

In this case, we say that the thresholds are **intrinsic**. An intrinsic threshold is a threshold that does not depend on the action a but only on the degree corresponding to the evaluation of this action. It can, therefore, be characterized by a positive valued function on E_i ($\eta_i^-(e)$ or $\eta_i^+(e)$ as introduced above).

Nonintrinsic thresholds generally arise when the evaluation $\gamma_i(a)$ uses data whose precision varies with the action considered. The following example illustrates these two types of thresholds.

Example 5: Media-Planning (from Section 8.1.5)

Consider first the elementary consequence total readership. The state indicator might be based on a survey technique. Sampling errors, difficulties respondents might have in understanding the survey questions, and imprecision in determining what constitutes a reader would be important factors to consider when determining the thresholds. Since there would be no reason to believe that these factors differ for various periodicals (if the readerships are expected to be similar), the thresholds are intrinsic. It would also seem plausible to state:

$$\eta_1^-(e) = \eta_1^+(e) = \eta_1(e).$$

Since sampling effects have a relatively greater impact on small readership levels than on large ones, it would not be realistic to consider $\eta_1(e)$ constant if the degree e (number of people in thousands) varies from a few units to several thousand. One might model $\eta_1(e)$ as being proportional to $\sqrt{e}$.

Now let a and a′ be two periodicals that have the same target correspondence:

$$\gamma_2(a) = \gamma_2(a') = \text{target readership/nontarget readership} = e \in E_2.$$

The precision of the state indicator γ_2 depends strongly on the size of the target readership or (see Section 8.1.5, ex. 5) on the strength of periodical in reaching the target population (denoted $\gamma_0(a)$). Under these conditions, the thresholds $\eta_2^-(a)$ and $\eta_2^+(a)$ cannot be intrinsic. Rather, they should be defined as functions of e and of $\gamma_0(a)$.

(continued in Section 10.1)

8.2.2.2 Dispersion thresholds and indicators

A dispersion threshold only makes sense in cases where the state indicator is a point indicator. One may wonder, however, what difference there would be between a point state indicator γ_i, supplemented by thresholds η_i^- and η_i^+, and a nonpoint indicator γ_i^* defined as:

$$\gamma_i^*(a) = [\gamma_i^-(a), \gamma_i^+(a)],$$

i.e., modeling the consequences of a as falling in a closed interval with lower and upper bounds of γ_i^- and γ_i^+, respectively. The difference is, in fact, small, but it does warrant clarification. To do so, it is convenient to place the idea of a nonpoint state indicator in a more general context. We do this with the following example.

Example 6: Research Projects (from Section 8.1.5)

Consider the integer scale based on the ratio of future costs to current costs. When the information provided by the principal investigator leads to a ratio between 5 and 8, for example, it would make sense to use an interval (a nonpoint state indicator) instead of a precise value (point state indicator) modulated by some indication of imprecision (as was done in Example 5 in Section 8.2.2.1). In this case, if no additional information was available, the interval would lead to $\gamma_i(a) = \{5, 6, 7, 8\}$, with all degrees being considered of the same precision.

Trying to "approximate" on a monetary scale M the level of external funding that the research project would possibly receive would not make sense either. Several funding levels could be envisioned, each corresponding to a specific hypothesis (e.g., concerning which group would support the project). Therefore, $\gamma_i(a)$ would cover several degrees of M, but not necessarily all the degrees of M between the lower and upper bounds of $\gamma_i(a)$. Also, since the various hypotheses would not be equally plausible, all the degrees of $\gamma_i(a)$ should not be considered on the same plane.

(continued in Section 8.2.3.1)

The notion of nonpoint state indicators is more general than that of thresholds. It also contains less information. The information that results when first determining a most plausible degree $\gamma_i(a)$ and then determining a lower and upper endpoint of the range surrounding it cannot be summarized as well by using a nonpoint indicator such as γ_i^* defined above. Nevertheless, the nonpoint state indicator still could be complemented by what we have called a dispersion indicator. In this case, we would try to associate a dispersion indicator with γ_i^* that depicts the information which this nonpoint indicator cannot convey. Although this may be a somewhat technical point, proceeding in this way has the advantage of providing a general means of defining the evaluation of an action, whether the state indicator is a point indicator or not (see Section 8.2.5).

Reasoning in terms of thresholds implies that all the degrees of the interval $\gamma_i^*(a)$ are not placed on the same plane. The degree $\gamma_i(a)$ receives a larger credibility than the other degrees, which in turn are not further differentiated from this point of view. This modulation of credibility can be described by a function $\delta_i^a(e)$ defined on the interval $\gamma_i^*(a)$. The function might be the following, for example (see Figure 8.2.1):

$$\delta_i^a(e) = \frac{2}{\eta_i^-(a) + \eta_i^+(a) + 2}, \text{ if } e = \gamma_i(a),$$

$$\frac{1}{\eta_i^-(a) + \eta_i^+(a) + 2}, \text{ if } e \in \gamma_i^*(a),\ e \neq \gamma_i(a). \quad \text{(r 8.2.1)}$$

By definition, this function is **a dispersion indicator representing negative and positive dispersion thresholds.**[12]

Figure 8.2.1: Dispersion indicator representing negative and positive dispersion thresholds

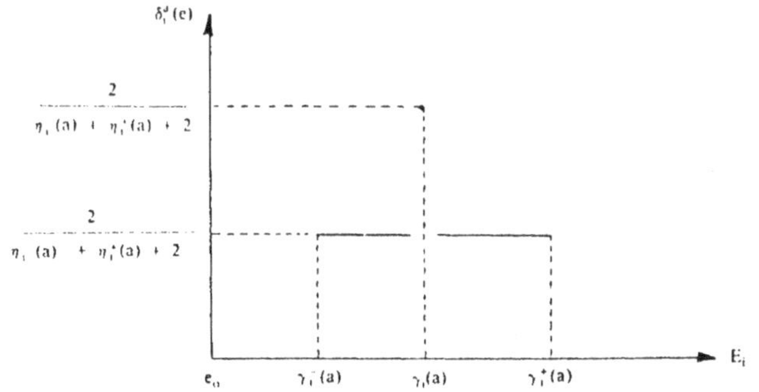

8.2.2.3 Properties of intrinsic dispersion thresholds

An intrinsic threshold (either positive or negative) is completely specified by a function $\eta(e)$ defined on the scale E (see Section 8.2.2.1); to simplify the notation, we omit the index i that indicates the dimension considered. We now show that this function has some restrictions. In passing from one degree to a higher degree, the positive threshold can increase without limit or remain constant, but it cannot decrease without limit. On the other hand, the negative threshold is limited in the manner in which it increases. Specifically, we present and justify the following result.

RESULT 8.2.1: A function $\eta(e)$ defined on a preference scale E must satisfy:

$$\frac{\eta(e') - \eta(e)}{e' - e} \leq 1;\ e,\ e' \in E$$

to represent a negative intrinsic dispersion threshold, and:

$$\frac{\eta(e') - \eta(e)}{e' - e} \geq 1;\ e,\ e' \in E$$

to represent a positive intrinsic dispersion threshold, where $e' - e$ is the difference in degrees for a numerical scale or the difference in the number of degrees that must be

---

[12] *translator's note: The original French version continues this discussion on Pages 206 and 207 at this point.*

traversed in passing from degree e to degree e′ for a nonnumerical scale. (In the latter case, e′ – e is considered positive if e < e′ and negative if e′ < e.)

Consider the case of a positive threshold. (The reasoning is similar in the case of a negative threshold.) Let $\eta(e)$ be a function for which there exists a pair of degrees contradicting Result 8.2.1. Letting e represent the lower degree and e′ the upper degree (which can be done without loss of generality), the hypothesis can be written:

$$\eta(e') - \eta(e) < e - e'.$$

If E is a numerical scale, this can be written:

$$e' + \eta(e') < e + \eta(e),$$

or:

$$\gamma^+(e') < \gamma^+(e)$$

where, as before, $\gamma^+(e)$ represents the degree $e + \eta(e)$. One can easily arrive at the same conclusion if E is nonnumerical by using the conventions defining the thresholds (see Section 8.2.2.1) and the difference e′ – e (see Result 8.2.1). But, the conclusion:

$$\gamma^+(e') < \gamma^+(e) \text{ with } e' > e,$$

is a contradiction. If (see Fig. 8.2.2) the margin of dispersion in increasing scale directions associated with degree e extends from e to $\gamma^+(e)$, then the degree $\gamma^+(e)$ must fall within the margin of dispersion in increasing scale direction of every degree e′ of the preference scale situated between e and $\gamma^+(e)$. Therefore, the hypothesis is not acceptable, and Result 8.2.1 holds in the case of upper intrinsic dispersion thresholds.

Figure 8.2.2: Function $\eta(e)$ that cannot represent an intrinsic positive threshold

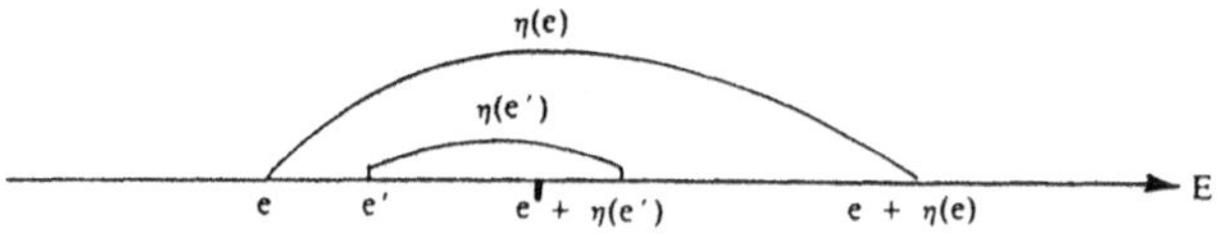

Note that the illustration presented in Figure 8.2.2 might well be encountered when considering two actions a and a′ in the case of nonintrinsic thresholds.

8.2.3 Modulated dispersion indicators (or modulation indicators)

Unless otherwise noted, the state indicators considered in this and the next sections are considered to be nonpoint indicators. In other words, $\gamma(a)$ (the index i being intentionally omitted except where needed for clarity) is a subset of the scale E. This subset is often, but not necessarily, an interval, where by interval we mean a subset I of E such that, if e and e′ are two degrees of I, then every degree e″ between e and e′ is also a degree of I.

In this subsection, we present the most basic category of dispersion indicators. In Section 8.2.1, we saw that imprecision, uncertainty, and inaccurate determination cause the different degrees of $\gamma(a)$ to be treated differently in many cases. Some degrees will be considered more important or more likely than others. The **modulated dispersion** indicator is used to model these distinctions in importance or likelihood as determined by the analyst.

Before formally defining this category of dispersion indicators in Section 8.2.3.5, we illustrate it and the approach taken to model it by considering four different cases. The cases are different in the information required to establish the indicator.

8.2.3.1 Dispersion factors determined from subjective opinions allowing distinctions in state importance or likelihood

Due to the complexity of the relevant phenomena or, perhaps, to a lack of time to investigate them properly, the analyst may not have objective information allowing him to distinguish among the different states of $\gamma(a)$ in terms of importance or likelihood. The following three examples show that important information can be obtained by considering subjective comparisons, intuitive orders of magnitude, and qualitative experience acquired in similar situations, and, more generally, by taking advantage of psycho-sensitive (senses and emotions) and integrative (intuitive and evocative) knowledge.

Example 2: Commuter Rail Line (from Section 6.1.3)

Suppose that, consistent with the objectives of Section 3.1, the analyst is interested in developing a dimension to model the effect of each variant on the travel time of a trip. Assume that he defines the states associated with this dimension in terms of the difference in trip times with respect to some reference variant operating under the same conditions as the variant considered. To define a dimension and a state indicator in this way, he must:

- characterize the reference variant, which may or may not belong to A. Suppose that a_{29} (see Fig. 5.2.3) is chosen as the reference variant. Since it is one of the poorest variants in terms of trip time, $\gamma(a)$ would be interpreted as a trip time savings.
- specify what is encompassed in trip time and fix the conditions for calculating it as a function of Section 5.2.2 hypotheses a)-e), which affect the definition of a variant.

As mentioned in Section 8.1.4, what appear as variations in trip times will strongly depend on how the trip times are calculated. This calculation will, in essence, define what is meant by trip time. Here, it would seem reasonable to consider average time, where the average is taken across the different origin-destination pairs, during peak and off-peak periods, and weighted by the corresponding amount of traffic generated. Without entering into details, it should be clear that when fixing the conditions for calculating a savings in trip time $\gamma(a)$, the analyst will face a wide range of options as to the factors to be considered and the values to be given to the various poorly estimated quantities involved in the calculations.

To avoid the numerous hypotheses that would have to be made to make $\gamma(a)$ a point indicator, suppose that the analyst will consider three calculations that will lead to the following degrees:

$\gamma^+(a)$: maximum time savings, which would result from particularly optimistic conditions;
$\gamma^0(a)$: median time savings, which would result from the most standard conditions possible;
$\gamma^-(a)$: minimum time savings, which would result from particularly pessimistic conditions.

The nonpoint state indicator that would follow in these conditions would be interval $\gamma(a) = [\gamma^-(a), \gamma^+(a)]$.

The analyst obviously believes that the degrees in this interval would not have equal weight or equal likelihood and could not, therefore, be used to evaluate the variant on the dimension being considered. Even though he cannot apply any empirical tests, everything leads the analyst to believe that the extreme values of the interval have lower weights or plausibility than values close to $\gamma^0(a)$. To portray the information embedded in this type of subjective opinion, a function $\delta^a(e)$ such as the following could be used:

$$\delta^a(e) = 0 \Leftrightarrow e \notin [\gamma^-(a), \gamma^+(a)];$$
$$\delta^a(e) \text{ is nondecreasing in the interval } [\gamma^-(a), \gamma^0(a)]; \qquad \text{(r 8.2.2)}$$
$$\delta^a(e) \text{ is nonincreasing in the interval } [\gamma^0(a), \gamma^+(a)].$$

It is not restrictive and often useful to force the values of the modulation indicator defined in (r 8.2.2) to sum to 1, i.e.:

$$\sum_{e \in \gamma(a)} \delta^a(e) = 1. \qquad \text{(r 8.2.3)}$$

Figure 8.2.3 portrays two examples of such modulation indicators. The analyst could, of course, envision others. Regardless of how he proceeds, some arbitrary assumptions will have to be made, and simplicity should, therefore, be considered of utmost importance. However, since the values of $\delta^a(e)$ only have an ordinal interpretation here, the impact of the arbitrary assumptions should be reduced.

Figure 8.2.3: Example of an ordinal modulation indicator defined on an interval based on a median value

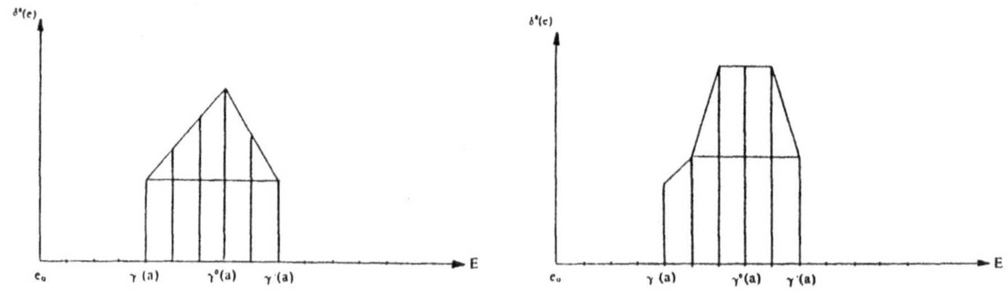

(continued in Section 8.2.4)

Example 4: Water Resource Planning (from Section 6.2.2)

Each of the short-term plans that constitute the actions of A imply the development of a coherent set of facilities. A date can be assigned to each action in A that would represent the probable time for beginning operations of the various facilities. The date t(a) that the last of these facilities would begin operations would be considered the installation date of action a.

Suppose that the analyst wishes to use an "installation date" dimension. He could use the point state indicator $\gamma(a) = t(a)$. However, this approach would not be satisfactory if, as we are assuming here, the uncertainty affecting the forecast of t(a) varies substantially from one action to another. Assume that based on the various facilities that need to be installed for each action, the analyst can determine an earliest installation date t′(a) and a latest installation date t″(a). The interval $\gamma(a) = [t'(a), t''(a)]$ would account for the variable fluctuations among actions. To each date in this interval, one could assign a positive number $\delta^a(e)$ that would increase with the perceived likelihood of state e. If the installation date t(a) seems realistic for the action a, $\delta^a(e)$ would be set at a maximum value for $e = t(a)$. If, on the other hand, delays of a few weeks or a few months seem inevitable, the maximum would correspond to a date later than t(a). In general, one would expect the time t′(a) to be among the least likely installation dates of the interval.

In this way, the analyst could use what are obviously subjective beliefs to order the various dates in the interval [t′(a), t″(a)] as a function of their likelihoods and represent the resulting order or preorder in a modulated indicator $\delta^a(e)$. If desirable, he could always multiply this ordinal indicator by an appropriate constant so that (r 8.2.3) holds.

(end of Example 4)[13]

Example 6: Research Projects (from Section 8.2.2.2)

Consider once again the state indicator $\gamma(a)$ that is meant to capture the level on a monetary scale M of the external funding that a project a might achieve. Unlike in the above examples, the degrees of $\gamma(a)$ do not necessarily form an interval on M. We saw in Section 8.2.2.2 that the monetary sums corresponding to the degrees $\gamma(a)$ were the result of certain hypotheses on the conditions and nature of the funding. The different hypotheses are not equally plausible, and the analyst could use his experience or interview those involved with the project to determine subjective probabilities for the various hypotheses. He could then determine nonzero numbers $\delta^a(e)$, $\forall\, e \in \gamma(a)$, that would correspond to the relative likelihood of obtaining funding equal to e. The ordinal modulation indicator defined in this way would satisfy (r 8.2.3).

(end of Example 6)[14]

8.2.3.2 Dispersion factors determined from objective observations allowing qualitative modulation of state importance or likelihood

In the cases presented in Section 8.2.3.1, there were no objective observations that allowed one to claim that $\delta^a(e)$ was larger than, equal to, or smaller than $\delta^a(e')$. In many cases, however, the analyst can take advantage of numerical information to determine values that indicate the importance or likelihood of the different states e of $\gamma(a)$. We

---

[13] *translator's note: Example 4 was continued and ended in Section 12.4 in the original French version.*

[14] *translator's note: Example 6 was continued in Section 9.3.4 and 11.2.2 in the original French version.*

refer to these cases below, as well as in Sections 8.2.3.3 and 8.2.3.4. Even though they are based on numerical information, the numbers $\delta^a(e)$ might only have a qualitative interpretation; e.g., e would have a greater impact or be more likely than e′ if a were implemented. As in Section 8.2.3.1, the dispersion indicator is again ordinal. We present two examples to illustrate this case.

Example 3: Agricultural Development (from Section 8.1.5)

Consider again the function $g_1(\underline{x})$ that provides the size of the trade deficit in agricultural products for the plan $\underline{x}$. This function was modeled as being linear in the components x_h of $\underline{x}$. Easily established formulas relate the coefficients of the function to the agricultural production, to the internal consumption of each product, and to the import and export prices of the products. The parameters of these relationships are not known with any confidence, however. Therefore, it would not be sufficient to use an "average" value for each parameter to specify the function $g_1(\underline{x})$ as a point state indicator.

Suppose that the analyst has estimates not only of an average value for each of the parameters, but also of upper and lower bounds on these values. He could use these values to calculate the minimum and maximum g_1 values for a given plan. These two numbers would specify an interval $\gamma_1(\underline{x})$ that would be a nonpoint state indicator. To modulate the relative likelihood of the different states, he could reason as follows.

The number $g_1(\underline{x})$ is obtained by adding together a large number of terms, each of which could be considered as being associated with some random error. The law of large numbers allows, at least in a qualitative way, the use of a Gaussian distribution to capture the likelihoods of the various values of this sum. The mean and standard deviation of this distribution could be determined either from an analytical solution or a numerical simulation. The numbers $\delta^{\underline{x}}(e)$ can then be easily determined for each of the states of $\gamma_1(\underline{x})$. The modulation indicator obtained in this way results in an "error analysis" that is justified by probabilistic reasoning. It would, therefore, seem legitimate to rank the likelihoods of the states of $\gamma_1(\underline{x})$ according to the values of $\delta^{\underline{x}}(e)$. These numbers could not, however, be considered objective probabilities.

(continued in Section 11.2.1)[15]

Let us present another example in a completely different setting. Consider the case in which each of k experts provides a point evaluation along a given dimension, $\forall$ a $\in$ A. Suppose that the different experts can be considered as being "of the same quality." Let $n^a(e)$ be the number of experts providing degree e in evaluating action a. It is normal to use a nonpoint state indicator $\gamma(e)$ that groups together all the degrees for which $n^a(e) \neq 0$ and to associate with it the modulation indicator defined by:

$$\delta^a(e) = \frac{n^a(e)}{k}.$$

Since the number k of experts is small, portraying the effect of the dispersion factors in this way would, of course, be qualitative. It would be necessary to consult a large number of experts, all being truly "of the same quality," for $\delta^a(e)$ to be interpreted as a probability.

---

[15] *translator's note: In the original French version, Example 3 was first continued in Section 9.3.4.*

8.2.3.3 Dispersion factors represented by distributions of nonrandom magnitudes allowing quantitative modulation of state importance or likelihood

This case and the next differ from the preceding two in that $\delta^a(e)$ possesses a cardinal, as well as an ordinal, interpretation. The example in Section 8.2.1d) illustrates this distinction. That example considered the consequence of the noise impact for a population distributed across a geographic area. The actions that might be associated with this impact (e.g., locations of highways, airports, or factories) differ from one another primarily in their physical location in the area. Let E be a scale where each degree specifies a noise impact level. As discussed before, it is not possible to describe the consequence of an action a by a unique state of E. It is necessary to modulate each of the noise impact levels affecting a nonnegligible fraction of the population. One might naturally use:

$$\delta^a(e) = \text{fraction of the population submitted to level e.}$$

This modulation indicator divides the population among the degrees of the scale. If one wished to group together some of these degrees (e.g., the two highest or the three lowest), one could add the $\delta^a(e)$'s corresponding to the appropriate degrees to obtain a new modulation indicator. In cases where the modulation indicator is ordinal, such an addition would have no theoretical justification, and the result would have no meaning.

8.2.3.4 Dispersion factors represented by probability distributions allowing quantitative modulation of state importance or likelihood

The example of the mayor's preferences in Section 7.2.1.2 illustrates this case. The motivation for the modulation indicator can be thought of as being due to a random phenomenon that influences those states of $\gamma(a)$ that will be obtained if action a is implemented. The indicator $\delta^a(e)$ is simply a probabilistic description of the phenomenon, and in this case, we speak of a **probabilistic modulation** indicator. This type of dispersion indicator is the only one considered in multi-attribute utility theory. This occasionally results in its being used without reference to the existence of any random phenomenon influencing the states of $\gamma(a)$ that will be obtained if a is implemented. When no objective probability distribution can be found, it is often possible to use techniques to encode subjective probability distributions (see Spetzler and Von Holstein, 1975; Wallsten and Budescu, 1983).

8.2.3.5 General form of modulation indicators: ordinal modulation and additive modulation

In the preceding cases, the same approach can be used to describe the consequences of an action on a dimension with a nonpoint state indicator. The importance and diversity of these cases justify the concept of dispersion indicators in a general approach.

DEFINITION 8.2.1: A ***modulated dispersion indicator (or modulation indicator)*** *is a nonnegative function* $\delta_i^a(e)$ *defined on part of a scale* E_i *associated with the action a that satisfies the following conditions:*

1) $e \in \gamma_i(a) \Rightarrow \delta_i^a(e) > 0$;

2) $\delta_i^a(e) = \delta_i^{a'}(e') \Leftrightarrow$ *the analyst attributes an equal importance or likelihood to degrees e and e′ when evaluating the consequences of a and a′, respectively, on dimension i;*

3) $\delta_i^a(e) < \delta_i^{a'}(e') \Leftrightarrow$ *the analyst attributes a lesser importance or smaller likelihood to the degree e than to the degree e′ when evaluating the consequences of a and a′, respectively, on dimension i.*

Definition 8.2.1 simply requires that the numbers $\delta_i^a(e)$ allow certain comparisons on the importance or likelihood of degrees e and e′ in evaluating either a single action or two distinct actions. It is, therefore, necessary that the way in which the numbers are defined does not vary from one action to another. When the manner in which the numbers are defined allows only simple comparisons (larger than, smaller than, equal to), the modulation indicator is said to be **purely ordinal** (see Sections 8.2.3.1 and 8.2.3.2). The following are two examples of manipulations of the δ's that require stronger than purely ordinal indicators.

Let $D \subset \gamma_i(a)$ and $D' \subset \gamma_i(a')$ be two subsets of degrees for which we would like to compare the importance or likelihood in order to evaluate the consequences of a and a′, respectively, on dimension i. One might be inclined to make the comparison by comparing the sums:

$$\sum_{e \in D} \delta_i^a(e) \text{ and } \sum_{e' \in D'} \delta_i^{a'}(e').$$

If the modulation indicator is defined in such a way that these sums and the comparisons of the sums have a meaning, then the modulation indicator is said to be **additive**. This is usually the case when δ has a probabilistic interpretation (see Section 8.2.3.4).

Calculating an average,

$$\mu(a) = \frac{\sum_{e \in \gamma(a)} \delta^a(e) \times e}{\sum_{e \in \gamma(a)} \delta^a(e)} \qquad \text{(r 8.2.4)}$$

is possible when the degrees of e are assigned numerical values. Such an average has little, if any, meaning if the modulation indicator is purely ordinal. It does have meaning when the modulation indicator is additive, however.

8.2.4 Referenced dispersion indicator

Even though there are many different possibilities resulting from the concept of modulated dispersion indicators, there are also many cases (see Section 8.2.1 h) where a lack of available information would make them insufficient or poorly suited to capture the dispersion factors. In these cases, a more complex formulation that depends on the specific case is required.

The following formulation provides a framework that is suitable to many problems. It is a generalization of the concept of modulation indicator and uses a set $\mathfrak{R}$, called a **dispersion referential**, whose definition and meaning can vary greatly from one problem to another. The set must be chosen such that it becomes possible to capture the effect of the dispersion factors when the problem is restricted to conditions corresponding to a single reference element $r \in \mathfrak{R}$. These factors are captured in a modulated dispersion indicator $\delta^a(e, r)$, and the set:

$$\{\delta^a(e, r)/r \in \mathfrak{R}\} = \delta^a(e, \mathfrak{R}) \qquad \text{(r 8.2.5)}$$

constitutes the dispersion indicator. In this case we speak of a **referenced dispersion indicator**. Without providing a more rigorous definition of this category of indicator, let us simply state that the number $\delta^a(e, r)$ must allow a comparison of the importance or likelihood of the states as implied by Definition 8.2.1 not only for the same reference element r, but also for two elements r and r′.

Example 2: Commuter Rail Line (from Section 8.2.3.1)

In Section 6.1.3 we emphasized the importance of considering various scenarios when comparing the variants. Indeed, evolutions over the next ten years in the residential, employment, and transportation structure of the area, or the addition of new connections with the rail system would affect the variants on several of the dimensions considered. Let S denote the set of scenarios to be considered.

First consider the case of one dimension – for example, suitability of the variant to urban development in the zones connected by the rail line – that would lead to a scale E with only a small number of degrees. Assume that the analyst can determine a single degree of E for each variant a if he considered a specific scenario s. Let $\gamma(a,s)$ denote this degree. The analyst, therefore, has developed a nonpoint state indicator: $\gamma(a) = \{\gamma(a,s)/s \in S\}$. By using this state indicator (i.e., the set of degrees E defined in this way), the relations between the degrees and the various scenarios would be obscured. When comparing a and a′, however, it might be useful for the analyst to be able to identify the degrees that correspond to the same scenario. A referenced dispersion indicator with $\mathfrak{R} = S$ would allow him to do so. One could simply define:

$$\delta^a(e,s) = \begin{cases} 1, & \text{if } e = \gamma(a, s); \\ 0, & \text{otherwise.} \end{cases}$$

This formulation may seen cumbersome in a simple case, but it allows one to consider the effects of the dispersion factor "future uncertainty" in a homogeneous fashion, as in the case of considering the set S of scenarios.

Consider another dimension, such as the trip time, for which things could be more complicated. In Section 8.2.3.1, we did not explicitly consider scenarios. It should be clear that even if the means of calculating the average savings in trip time – i.e., of determining $\gamma^-(a)$, $\gamma^0(a)$, and $\gamma^+(a)$ – is essentially the same, whatever scenario s is considered, a certain amount of data used in the calculation (e.g., the proportion of trips made between the various origin-destination pairs) may still vary drastically from one scenario to another. Therefore, one would need to calculate the numbers $\gamma^-(a, s)$, $\gamma^0(a, s)$, $\gamma^+(a, s)$. One could again use these numbers to define a modulated dispersion indicator $\delta^a(e, s)$. If the formula used to determine this function is independent of s, once the three values $\gamma^-(a, s)$, $\gamma^0(a, s)$, $\gamma^+(a, s)$ are fixed, then the set $\{\delta^a(e,s)/s \in S\}$ can be considered a referenced dispersion indicator.

(continued in Section 9.2.2.1)

Example 8: Airport Operations (from Section 6.2.2)

In the second phase of this study, each action consists of a few initial fragmented actions that together form a coherent plan. Let E be a scale associated the dimension, "passenger reaction," with degrees: hostile reaction, somewhat negative reaction, neutral reaction, somewhat positive reaction, very favorable reaction. The composite nature of the actions and the heterogeneity in the passenger population (tourists, business travelers, regular passengers, occasional passengers, ...) would necessitate the use of a nonpoint state indicator for this dimension. Questioning the passengers directly could lead to the percentage of the population having reactions corresponding to each of the five levels of E for each plan. This approach would be expensive and time consuming, however, and the analyst rejects it in favor of discussions with three individuals whose responsibilities make them knowledgeable of the potential passenger reactions, but from three different perspectives. Each of these three individuals segments the passenger population in different ways and examines the potential passenger reactions in each segment to the fragment of each plan. In this way, each establishes a function $\delta^a(e)$ for each action a and each of the five degrees e. This function expresses the fraction of the total passenger population that would have a level of satisfaction corresponding to e if the action a were implemented. The three different modulation indicators – one for each of the three individuals – would be a referenced modulation indicator.

(continued in Section 12.3)

8.2.5 Evaluating an action: Principles of clarity, universality, and reliability

By evaluation of an action, we mean the model $\Gamma(A)$ that formally combines all the information that could be useful in describing the consequence cloud of any action in A. When applied to a particular action a in any of the previously discussed cases, this model could be written in the following general form:

$$\Gamma(a) = \{\gamma_i(a),\ \delta_i^a(e)/i = 1, ..., \overline{n}\}. \qquad \text{(r 8.2.6)}$$

The γ_i's and δ_i^a's are called **components of Γ(A)**.

Actually, the set of components $\delta_i^a(e)$ contains all the information, and one could avoid the use of the γ_i components. The information contained in the γ_i's is usually the most important, however, and we prefer to make this explicit in a separate set of components and use the set of δ_i^a's as complementary information. With this convention, the component δ_i^a might be omitted in some cases – e.g., cases with point indicators and no thresholds, cases with nonpoint indicators but with identical importance or likelihood across all states.

In some cases, we will be interested in modeling a partial cloud $\upsilon_g(A)$ of $\upsilon(A)$, where g denotes a category of elementary consequences. The submodel of $\Gamma(A)$ corresponding to $\upsilon_g(A)$ will be denoted $\Gamma_g(A)$.

The model $\Gamma(A)$, therefore, seems to be a complete and unbiased (i.e., without arguable hypotheses) representation of the class of phenomena formed by the consequences of any action in A. Moreover, this representation is general for any actor Z whose preferences are important to the stage of the analysis considered. The methodology presented in this chapter to assist in building such a model has been progressively conceptualized and tested so as to accommodate the three following principles:

a) Principle of clarity

The components of Γ(A) must directly describe the various consequences as the actors see them or are apt to understand them with respect to the preference scales.

b) Principle of universality

The components of Γ(A) must be associated with dimensions reflecting basic preference judgments about the actions of A that are held by all the actors.

c) Principle of reliability

The model Γ(A) must explicitly express the reliability (level of precision, meaning, validity) associated with the most important components and do so as a function of the action a considered.

These principles emphasize the relative importance given to the modeling of subjective or hard-to-quantify effects. Contrary to what we have often observed, we believe that it is preferable to have an opinion based on imperfect information examined in an exhaustive manner than one based on exact calculations that do not account for certain essential factors (e.g., instrumental biases, see Section 2.2.6).

Chapter 9

COMPARING ACTIONS AND DEVELOPING CRITERIA

SUMMARY

In Section 9.1.1 we review the different meanings of the term "criterion" and mention that criterion and criterion function are usually synonymous in the field of decision aiding. We then define (Def. 9.1.1) and generalize this concept in Section 9.1.2 and add five remarks.

We address the development of criteria from action consequences that are modeled by state and dispersion indicators along various dimensions in Section 9.2. In Section 9.2.1 we discuss the case when criterion g is associated with a single dimension i that leads to a point evaluation. In this case, we show (Res. 9.2.1) that developing such a criterion leads to a state indicator encoding (Def. 9.2.1) and describe several examples. In Section 9.2.2, we discuss the case when the criterion g is associated with a single dimension i that leads to a nonpoint evaluation. Here, we distinguish between two cases. In the first case (Section 9.2.2.1), g is the only criterion that affects the evaluation along dimension i, and we say that there is point reduction in the dimension. We discuss and illustrate the principal point reduction techniques, especially the technique based on utility theory. In the second case (Section 9.2.2.2), g is not the only criterion that affects the evaluation along dimension i, and we say that the criteria split dimension i. We discuss the reasons for such splitting. Finally, in Section 9.2.3 we consider the case when a criterion g_I affects all the dimensions in a subset I. When g_I is conceived so as to contain all the evaluation information contained in this subset of dimensions, we say that there is subaggregation of these dimensions. We explore when developing such criteria makes sense and clarify terminology.

In Section 9.3 we investigate the limits of using the values of a criterion to form indifference or strict preference relations over pairs of actions. This leads us to propose the notion of discriminating power of a criterion in Section 9.3.1. We provide an operational meaning to this notion in Section 9.3.2 when we introduce the concepts of indifference and preference thresholds. In Section 9.3.3 we define the important concept of pseudo-criterion as a criterion function to which discrimination thresholds are added (Def. 9.3.2). We introduce specific cases where at least one of the two thresholds is empty (Def. 9.3.3). The structures of the resulting systems of preference relations correspond to structures already seen in Chapter 7 (Res. 9.3.1). We finish this subsection by discussing practical ways to determine discrimination thresholds.

In Section 9.4, we examine when the difference $g(a') - g(a)$ can be used to reflect the qualitative importance of the difference in actions a′ and a according to the criterion g. After motivating the general problem in Section 9.4.1, we present the two important definitions of gradation (Def. 9.4.1) and gradable criterion (Def. 9.4.2) in Section 9.4.2. We devote Sections 9.4.3 and 9.4.4 to criteria that can be called measures. In Section 9.4.3, we present the basic definition (Def. 9.4.3), provide examples, and discuss properties (Res. 9.4.2). In Section 9.4.4, we discuss the special case of the von Neumann-Morgenstern expected utility criterion. We specify the axiomatic basis (Axioms 9.1-9.4) that defines this criterion up to a positive affine transformation (Res. 9.4.3) and discuss the conditions under which it is a measure. These conditions are closely related (Axiom 9.5 or 9.6) to those that allow differences in criterion values to reflect importance.

9.1 THE CONCEPT OF CRITERION

In Chapter 8, we presented a methodology designed to describe and formalize the fundamental elements that the different actors use to develop, justify, and transform preferences relative to potential actions. In Chapter 7, we discussed the basic concepts required to construct a formal representation of these preferences. We now are ready to address the central idea of criterion. This concept is often associated with that of a function.

What does the word criterion represent, and what is a criterion function? How does this new idea help the decision aiding task, and what is its relation to the preceding concepts of dimensions, state indicators, and dispersion indicators? The first two sections address these questions. The last two sections allow the reader to appreciate the richness and precise structure (in the sense of Section 7.2) of the preference model based on the criterion function. The way in which one or several criteria can help synthesize the information contained in $\Gamma(A)$ will be discussed in Chapter 10.

9.1.1 Criteria and functions: General remarks

The word criterion commonly signifies, "that which serves as a basis of a judgment: style is not the only criterion for judging the value of an accomplishment" (translated from the Robert dictionary). This is the usual sense of the word criterion in operations research, decision theory, and, more generally, in decision aiding. In this context, the judgments that the criterion must help establish are essentially the preference judgments related to the decision. This is how we use the word criterion in this book.

We note that in management science, the word criterion often has a less restrictive meaning, one which is closer to its etymology (the Greek origin is related to an ability to discern), and somewhat related to that used in philosophy[1]: "characteristic, sign that allows a distinction of a thing, a notion, an assessment of an object." We shall avoid using the word criterion to mean a simple characteristic or sign that serves as a basis of an estimation or assessment whose only purpose is to discern or distinguish in the way in which hair color, age, or socioeconomic categories can distinguish among individuals in a population. In this case, we shall speak of attribute. For us, an attribute will be a criterion only when it can serve as a basis for a preference judgment.

As such, in the highway location example (see Table 7.1.2), committee Z can speak of a criterion of investment cost, a criterion of noise pollution, or a criterion of damage to forested areas. The notions of criterion and dimension are closely related, but the distinction is worth noting. The investment cost, noise pollution, and damage consequences mentioned above could be integrated into a single criterion that would no longer constitute a dimension in the sense of Definition 8.1.4. Conversely, the first two columns of Table 7.1.2 can be considered as two distinct criteria associated with the dimension of noise pollution.

---

[1] *translator's note: Translated from Robert, 1968.*

Translating from Optner (1968), "the use of a criterion constitutes a preference test." It is a test in the same way that a constraint can be considered a test of feasibility. This test may require the use of all of the consequences, i.e., the entire $\Gamma(A)$ discussed in Section 8.2.5; or, it could be **restricted to certain components** of the general model of $\upsilon(A)$. The criterion, then, is a model that uses some or all of the information explicitly represented in $\Gamma(A)$ to form the basis of a relative judgment as to whether an action is better or worse than some other action or an absolute judgment as to whether an action is better or worse than some reference actions that are intended to represent what is good or bad.

Although in practice, the formality of models leading to preference judgments may vary, the term criterion is usually reserved for models that possess the properties of a real-valued function defined on A.[2] This is why **we shall use criterion and criterion function interchangeably**.

Example 11: Product Composition (from Section 6.1.1)

Let $\underline{x}$ and $\underline{x}'$ be two rubber specifications considered to be consistent with the initial request. The technical division is using the following vectors to guide its choice:

$$\underline{y} = (p_0(\underline{x}), \ldots, p_{12}(\underline{x}));$$
$$\underline{y}' = (p_0(\underline{x}'), \ldots, p_{12}(\underline{x}')).$$

(Recall that the function p_k, for $k \neq 0$, gives the performance level for a specification relative to the kth property and that p_0 gives the total cost of the specification.)

All the parties involved speak of "minimizing cost," "maximizing the fracture stress," "optimizing the heat build-up," and, by doing so, refer to criteria. Each criterion is associated with what we shall call a significance axis, an axis that allows a comparison on the basis of francs, kilograms per centimeter cubed, degrees Celsius, ... The significance axes of such criteria are easily linked to dimensions associated with the elementary consequences of cost and the performance levels of the properties considered.[3]

When satisfaction varies monotonically with the value of $p_k(\underline{x})$, such as is the case of cost ($k = 0$), the attribute defines an elementary consequence whose states can be ordered in such a way as to form a dimension. The function p_k would form its point state indicator. It may not be as straightforward in the case of the physical properties,

---

[2] *Recall that a real-valued function g defined on a set A is a procedure (e.g., a formula, table) that associates a unique real number g(a) to each element a of A.*

[3] *If one tried to aggregate the three criteria mentioned into a single criterion, this aggregated criterion would no longer express a concrete unit. Its significance axis would have to be considered in more abstract terms, such as utility, and it would no longer be possible to associate the significance axis with a dimension of elementary consequences.*

however. It is possible that the satisfaction of the party requesting the rubber is maximized at some value $\hat{y}_k$, which might then be considered as an objective. The increasing or decreasing values of $p_k(\underline{x})$ would not define a dimension in this case, and one would have to use some construct like the differences from the objective (Section 8.1.3). The state indicator might be $|y_k - \hat{y}_k|$ or something more complicated if positive and negative deviations from the objective are not equally valued. In either case, with or without most preferred levels $\hat{y}_k$ (objectives), the state indicator could serve as a criterion.

As mentioned in Section 5.2 (see (r 5.2.1)), the various properties and the cost can be important in defining A. When there is more than one criterion, it is possible that a component of $\underline{y}$ acts both as a constraint on the definition of A and as a criterion. It is also possible that a criterion of p_k becomes unimportant as soon as $p_k(\underline{x}) \geq y_k^*$, where y_k^* is a performance level such that levels above it (considering nondecreasing preferences in y_k) are not considered to be any better than y_k^*.

Table 9.1.1: Example relating to three criteria

Actions	External Heat Resistance in minutes	Fracture stress in kg/cm²	Heat Build-up in Degrees Centigrade[4]
a^1	13	260	50
a^2	12	500	40
a^3	20	230	40
a^4	20	500	55
a^5	20	200	47
a^6	15	500	52

When the model has more than one criterion, the technical service must obtain more information than that contained in each criterion's significance axis to compare two specifications $\underline{x}$ and $\underline{x}'$. If not, it would have to admit situations of incomparability when comparing many pairs of actions – e.g., all pairs except a^3 and a^5 in Table 9.1.1, where the performance levels considered are the external heat resistance, fracture stress, and heat build-up. The point is that the individual criteria that capture separable, important, but restricted categories of consequences are not rich enough for making comprehensive comparisons, and the preference modelling covered in this second part of the methodology (see Section 4.2.2) is usually not sufficient.

(end of Example 11)[5]

[4] *Preference decreases with increasing values.*

[5] *translator's note: Example 11 was continued and ended in Section 11.4.2 in the original, French version.*

9.1.2 Definition and comments

The concept of criterion function is a traditional one. As an example, Fishburn (1978) says "...a criterion function usually indicates a real-valued function on X that directly reflects the worth or value of the elements in X according to some criterion or objective. These functions are also referred to as objective functions. Unlike attribute mappings, which usually describe objective characteristics of alternatives or consequences, criterion functions often represent subjective values on a more or less arbitrary scale. However, values of criterion functions may have objective content such as net profits, test scores, times until completion, payback periods, expected values and market shares."

DEFINITION 9.1.1: A real-valued function g defined on A is a ***criterion function or a criterion*** *capturing the subcloud of consequences $\upsilon_g(A)$ for actor Z if:*

1) *the number g(a) is obtained if and only if an evaluation $\Gamma_g(a)$ of $\upsilon_g(a)$ is available; the model $\Gamma_g(a)$ that provides this evaluation is called the support of the criterion function g;*
2) *actor Z, or the analyst judging in her name, recognizes the existence of a significance axis on which any two potential actions a and a′ can be compared according to the consequences covered by $\upsilon_g(A)$ alone, and she accepts the following as a model of this comparison:*
   $g(a') \geq g(a) \Rightarrow a' \; S_g \; a,$
   where S_g denotes an outranking relation restricted to the significance axis of criterion g (where all the other aspects of consequences not modeled in the support of g are ignored[6]).

Before illustrating this definition, we present the following remarks.

a) The sign of the inequality in condition 2 represents the convention that the better the action according to criterion g, the larger its value g(a). If the consequences are such that the opposite is true (i.e., actions decrease in value as g(a) increases), such as with a cost, changing g to –g would allow us to keep the conventions developed. We add here, however, that, in order to avoid using negative numbers we shall sometimes (as in Table 9.1.1) speak of a criterion g even though:

$$g(a') \leq g(a) \Rightarrow a' \; S_g \; a.$$

b) We use the weakest relation S_g in condition 2 of Definition 9.1.1, rather than one, or several, that model tighter results (see (r 9.1.1) below) for the following reason. Obtaining two numbers g(a) and g(a′) assumes both a model and what we, somewhat improperly, call data. The model defines criterion g (see Section 9.2), and the data

---

6 *Some readers, especially those familiar with multiattribute utility theory, may object to this phrase, questioning its implications for independence hypotheses that need to be verified. Since we are only concerned with specific problems related to developing criteria in this chapter, we ask these readers to ignore such difficulties until we address the issues of developing a coherent criterion family in Chapter 10.*

correspond to the values of the components of a and a′ included in the support of g (see Section 8.2.5). The numbers g(a) and g(a′) result from simplifications that may be somewhat arbitrary, but that are necessary to overcome inaccurate determinations and allow a model of g. The imprecision, uncertainty, and inaccurate determination that occur throughout the analysis will also affect the values given to the data. Asserting a tighter result than an outranking relation would, therefore, only seem to be justified in special circumstances, such as the following:

- One states conditions that capture the degree of importance of the difference g(a′) − g(a): one could, for example, define threshold functions that specify when S_g can be replaced by I_g, Q_g, P_g (see Section 9.3, below).

- One no longer considers real actions, but **ideal actions**, where comparisons **correspond exactly** to those of the numbers g(a) and g(a′). Considering such hypothetical situations, by definition, amounts to accepting that any approximation, inaccurate determination, or arbitrary element that could affect the values of criterion g is assumed away. In this hypothetical world of ideal actions all the "data" are ideal, and there are, therefore, no errors in the model g. For such ideal actions, the final implication of Definition 9.1.1 can be replaced by:

$$g(a') > g(a) \Rightarrow a' \succ_g a,$$
$$g(a') = g(a) \Rightarrow a' \; I_g \; a,$$

(where $\succ_g$ and I_g denote, respectively, preference and indifference according to the significance axis g, where all other aspects of consequence not modeled in the support of g are ignored).

c) In addition to being often somewhat incomplete (omitting Condition 1 in Definition 9.1.1), the definitions of criterion found in the traditional literature substitute the following pair of implications for the second condition:

$$g(a') > g(a) \Leftrightarrow a' \; P_g \; a,$$
$$g(a') = g(a) \Leftrightarrow a' \; I_g \; a. \qquad \text{(r 9.1.1)}$$

(where P_g and I_g denote, respectively, strict preference and indifference according to the significance axis g, where all other aspects of consequences not modeled in the support of g are ignored). Using these implications in Definition 9.1.1 limits its applicability to one special case (see Section 9.3). Note that the second implication of (r 9.1.1) seems restrictive, even when considering ideal actions, since every nonzero difference, no matter how small, results in a strict preference.

d) The subcloud of consequence $\upsilon_g(A)$ captured by a criterion g enters the function g only through its representation in the model $\Gamma_g(a)$. When an evaluation model $\Gamma(A)$ is defined for the entire $\upsilon(A)$, the support $\Gamma_g(A)$ is, quite naturally, the part of $\Gamma(A)$ restricted to $\upsilon_g(A)$. The link between the function g and its support may be quite simple, as when each point state indicator defines a criterion (see Section 9.2.1), or it may be much more complex (see Sections 9.2.2 and 9.2.3).

e) When referring to a criterion such as the present value of profits, the possibility of some event occurring, the noise level, the degree of obtaining an objective, the suitability of some policy, ..., the actor or the analyst is not only considering the concrete contents of the elements modeled in the criterion support, but also the point of view she or he is taking to give meaning to the significance axis of the criterion. Even when the concrete contents of the consequences considered or the exact nature of the significance axis remain relatively imprecise, it is the combination of the concrete contents, the nature of the significance axes, and the point of view adopted that allow one to give a name to the criterion g. When the function is defined beforehand, perhaps for convenience, one must be careful that the name given to it does not lead to false impressions.

   An actor or an analyst may consider several significance axes as being pertinent to a given subcloud $\upsilon_g(A)$ and evaluation model $\Gamma_g(a)$. (Examples are given in Section 9.2.2.2.) Several criteria will then have the same support, and their names will have to reflect the distinction among different points of view or the representation systems underlying them.

f) When $\upsilon_g(A) = \upsilon(A)$ (therefore, $\Gamma_g(A)$ can be considered the same as $\Gamma(A)$) and when the significance axis is not specific to a point of view or to a special representation system, we shall speak of a **comprehensive comparison**, and the associated relations as **comprehensive outranking, comprehensive preference, comprehensive indifference**, ... Otherwise, we shall speak of **restricted comparison, restricted outranking, restricted preference, restricted indifference, ..., along the significance axis of the criterion.**

If the criterion is to capture all of $\upsilon_g(A)$ in a significance axis characterizing this subcloud with no other significance axes in the same subcloud, we shall call the axis a **totalitarian** significance axis, and speak of comparison, preference, indifference, ... **restricted to subcloud** $\upsilon_g(A)$. We add that such a restricted comparison will be a comprehensive one whenever the pair of actions considered differs only in those aspects of the consequences captured in $\upsilon_g(A)$. Thus, if criterion g is constructed relative to a totalitarian significance axis, one could substitute S (comprehensive outranking) for S_g in Definition 9.1.1 whenever[7] a and a′ are such that:

$$\bar{\Gamma}_g(a) = \bar{\Gamma}_g(a'),$$

where $\bar{\Gamma}_g(a)$ denotes the set of components of $\Gamma(a)$ not in $\Gamma_g(a)$.

In this chapter we address the decision aiding issue of developing a criterion that faithfully reflects in the significance axis an actor's preferences for those aspects of the

[7] *See Def. 9.1.1 and the corresponding endnote.*

consequences considered. In practice, this only has meaning when reasoning on the basis of a certain number of actual actions. Nevertheless, the significance axis of a criterion g must allow comparisons not only among these actual actions, but among all other realistic, even dummy actions (see Section 5.1.1) that may arise. This is why, when beginning to think about g as well as when discussing its properties, one must reason on the basis of a set A that contains both the actual actions implied in the phase of the study considered and all the dummy actions consistent with consequences that could arise if they corresponded to realistic decision contexts. Specifically, every possible value of $\gamma_i(a)$ and δ_i^a must be obtained for at least one action a of a set defined in this way. This leads to a set of possible values for g(a), which we shall denote g(A). From now on, when we refer to a set of potential actions A without further specification, the set must be seen as being **saturated with realistic dummy actions**. More specifically, consistent with the explanations given in remark b) above, we can in a certain sense consider the actions of A as ideal actions for which there is no reason to determine whether or not they are realistic.

9.2 CONSTRUCTING CRITERIA FROM CONSEQUENCES

How can we form criteria from action consequences that are modeled in terms of point state indicators and dispersion indicators in several dimensions? This second section treats the principal processes that we can use to construct criteria. We do not discuss criteria that should be used to account for the set of consequences "as conveniently as possible" until the next chapter, however.

The first two parts of this section deal with those cases where the significance axis of criterion g uses only one of the dimensions of the spectrum, while the third part treats the other cases.

In the following, i denotes a dimension of the spectrum $\overline{v}$, and γ_i and δ_i are the state and dispersion indicators that define the evaluation on this dimension (Section 8.2.5).

9.2.1 Criterion function with one dimension and a point state indicator

In Section 9.1.1 Example 11 demonstrated that, when it is a real valued function, γ_i can be selected as the criterion function associated with dimension i. This natural use of a real-valued point state indicator as a criterion function is always possible when $E_i \subset \mathbb{R}$. Examples can be found in Tables 7.1.2 and 7.1.3.

When, on the other hand, the degrees of E_i are defined in a qualitative manner (i.e., $E_i \not\subset \mathbb{R}$), defining a criterion function on the basis of the state indicator seems less straightforward and even somewhat arbitrary, since $\gamma_i(a)$ is not a real number. The dimensions of comfort and safety in Table 7.1.1 are examples of this case.

We shall see that there is really no fundamental difference between those cases when the degrees of the scale are contained in $\mathbb{R}$ and those when they are not. Despite any associated connotations, the means used to denote degrees are primarily a way to communicate the characteristics of the scale, that is, the complete order of the degrees reflecting preference. Therefore, the degrees can be denoted in various ways without altering the basic characteristics of the scale, and it should always be possible to represent degrees by numbers.

DEFINITION 9.2.1: Let E_i be a set for which there exists a complete order denoted $>^i$. We shall call an ***encoding*** *of the scale E_i any real valued function χ defined on E_i that has the following property:*
$e >^i e' \Leftrightarrow \chi(e) > \chi(e')$.

To illustrate this definition, first consider a qualitative scale E_i whose degrees are numbered according to the order of preference. The discomfort and functional breakdown scales introduced in Section 8.1.3 can serve as examples. The functions χ_1 and χ_2:

$$\chi_1(e_h) = h, \ \chi_2(e_h) = 2^h, \ h = 0, 1, \ldots \qquad \text{(r 9.2.1)}$$

would be two distinct encodings of E_i.

Let us assume now that E_i is a quantitative scale, either discrete or continuous. The identity function that associates the number x with each degree of the same number x is an encoding that we shall call an identity encoding. To define a criterion function on the basis of a real valued state indicator, the identity encoding can be transformed into a different encoding. For example, if E_i is a monetary scale (Section 8.1.3), or a time, stress, or temperature scale as in Example 11 (see Section 9.1.1), the analyst might have some reason to use encodings such as χ'_1 or χ'_2 defined as:

$$\chi'_1(x) = a \cdot x + b, \text{ with } a > 0,$$
$$\chi'_2(x) = a^x, \text{ with } a > 1. \qquad \text{(r 9.2.2)}$$

Let us mention two other useful encodings when faced with ratios (see Table 7.1.3) or risk scales (see Section 8.1.2):

$$\chi_3(x) = \frac{x}{1 - x}, \text{ with } 0 \leq x < 1;$$

$$\chi_4(x) = \mathrm{Log}\left(\frac{1}{1 - x}\right), \text{ with } 0 \leq x < 1. \qquad \text{(r 9.2.3)}$$

The reader can easily verify that these are encodings in the sense of Definition 9.2.1.

The above discussion leads us to state the following result:

RESULT 9.2.1: Let i be a dimension on which γ_i is a point state indicator. Whatever the encoding χ of the scale E_i, $g_i^\chi(a) = \chi[\gamma_i(a)]$ is a criterion function that accounts for preferences restricted to the elementary consequence corresponding to dimension i. E_i can be considered the significance axis of $g_i^\chi(a)$, and $\gamma_i(a)$ can be considered its support; $g_i^\chi(a)$ will be called a natural criterion function associated with E_i by the encoding χ.

To justify this result, one can reconsider Definition 9.1.1 point by point in light of Definition 9.2.1.

Note that the significance axis to which the criterion function $g_i^\chi(a)$ refers can only be a totalitarian one (Section 9.1.2,f) if the state indicator is a point indicator and the subcloud of consequences reduces to a single dimension. Even so, one may wonder if this criterion function accounts for all of the information contained in the evaluation along dimension i. This is indeed the case when the positive and negative dispersion thresholds are both zero. When they are not, this information is not accounted for by the function g, at least at this stage.

In fact, we shall see in Section 9.3 that distinctions among the preference situations covered by the outranking relation S_g (Def. 9.1.1) make it relatively easy to account for the extra information provided by the dispersion thresholds. This will lead us to introduce other thresholds called discrimination thresholds. Finally, in Section 9.4 we shall see the possibilities for modeling preference differences among degrees that result from the choice of encoding.

9.2.2 Criteria with one dimension and a nonpoint state indicator

Unlike the case discussed above, we now consider the elementary consequence c_i (corresponding to dimension i) to be such that a unique criterion cannot account for all the information it contains. The analyst can either:

- use a single criterion to model c_i at the price of losing a great deal of information with respect to $[\gamma_i, \delta_i]$; in general, this is like substituting a more complex point indicator for the nonpoint state indicator; the criterion constructed in these cases is, therefore, called a **point reduction criterion on dimension i** (see Section 9.2.2.1);

- use several criteria to portray the complexity of c_i better; in these cases the criteria are called **splitting criteria of dimension i** (see Section 9.2.2.2).

Whether it results from the first or second approach, a criterion g attempts to capture the subcloud $\upsilon_i(A)$ forming c_i. When there is splitting, each criterion used is necessary to account for some specific aspect of $\upsilon_i(A)$ while rounding out the other criteria having the same support; therefore, none of them has a totalitarian significance axis (see Section 9.1.2,f). On the other hand, when there is point reduction, criterion g is the only one used to account for $\upsilon_i(A)$, and the claim is that it captures this subcloud completely. Thus, as in the point indicator case, its axis is totalitarian. Therefore, the significance of a given function g can depend on whether g is a splitting or a point reduction criterion.

In either case, the function g must have the form:

$$g(a) = G[\gamma_i(a),\ \delta_i^a(e)]. \qquad \text{(r 9.2.4)}$$

Unless stated otherwise, $\delta_i^a(e)$ will be considered a modulator (Section 8.2.3), where the point reduction or splitting criteria are, in some sense, intended to allow comparisons of distributions, especially probability distributions. This is exactly their role when comparing ideal actions.

9.2.2.1 Point reduction on the dimension

We now review the most common forms of the operator G by grouping them in the two families discussed in a) and b). The concept of a point equivalent is presented in c).

a) Point reduction based on an average or a global mass

a1) Examples and basic formulas

When thinking of using an average to summarize dispersed elements along a dimension, the case of a probability distribution that quantitatively modulates the importance or likelihood of states (see Section 8.2.3.4) comes to mind. Thus, if e denotes a discrete measure of the annual volume of water liable to be collected in the reservoir of a dam a, and if $\delta_i^a(e)$ is the probability that this volume is exactly e in one of the next few years, then the average annual volume of water can be used as a point reduction criterion. It can be written:

$$g(a) = \sum_{e \in \gamma_i(a)} e \cdot \delta_i^a(e).$$

In this case, the operator G corresponds to a mathematical expectation that is calculated from the natural values of the degrees. As we see in a2) below, it will be preferable to perform the calculation based on an encoding of the degrees.

Let us now consider the case where $\delta_i^a(e)$ represents the fraction of population submitted to a noise level e associated with action a (see Section 8.2.3.3). Criteria in the family,

$$g^v(a) = \sum_{e \in \gamma_i(a)} v(e) \cdot \delta_i^a(e), \qquad \text{(r 9.2.5)}$$

with v(e) a function defined on E_i, constitute point reduction criteria. Their significance axes evidently depend on the choice of the function v. For example, if v(e) is an encoding of E_i intended to represent the amount of money that would have to be paid to each individual submitted to noise level e associated with action a to compensate for this negative effect, then $g^v(a)$ represents the average cost of the noise effect due to action a. If, on the other hand, v(e) is 1 for every degree greater than a threshold ê and

0 for all the others (strictly speaking, v is no longer an encoding), then $g^v(a)$ represents the fraction of the population exposed to a noise level greater than ê, given action a.

Analysts in the transport domain use the same type of criterion function when introducing a criterion "time saved by users" (e.g., in locating a highway, designing a mass transit line, planning an airport). This criterion has the form of (r 9.2.5) with $\delta_i^a(e)$ representing the number of passengers for whom action a leads to a "time savings e" and v(e) being the level of time e. The analyst who states v(e) = e is using the total mass of time "saved" as a point reduction criterion. That is, he is considering 30,000 people saving one tenth of a minute each as equivalent to 300 people saving ten minutes each. He could express different forms of equivalence with different v functions.[8]

a2) Basic remarks and a first look at utility theory

A point reduction criterion based on an average or a global mass forms the product of v(e) and the value given to the dispersion index for that degree and then the sum of these products over all degrees. That is, to consider this means of point reduction:

- these operations of multiplication and addition must have meaning with respect to the nature of the information represented by the numbers δ; this leads us back to the discussion at the end of Section 8.2.3.5 and shows that this means of point reduction is almost always devoid of any significance when δ is purely ordinal;
- the concrete implications of the calculations must be sufficiently clear to determine whether the preference model that stems from the criterion built in such a way (see Def. 9.1.1) corresponds to a realistic totalitarian significance axis (see the time savings example above).

How to choose the form of v(e), which is often called a "value" or "utility" function, is not always obvious. This function would not necessarily be strictly monotonic in E_i and, therefore, not necessarily an encoding of E_i. When faced with more complicated cases than those presented above, the analyst will often propose that the number v(e) is an abstract utility associated with degree e and that $g^v(a)$ is a comprehensive utility that aggregates the utilities of the various possibilities related to the dispersion. This does not address the difficulties inherent in the second requirement stated above, however, even if the dispersion indicator represents a probability or the size of a population.

To overcome this type of difficulty, von Neumann and Morgenstern (1967) developed their utility theory to cover the case where δ is a probability distribution. As we shall see in Section 9.4.4, the theory makes the necessary and sufficient conditions explicit for a function g^v to define a preference model according to relation (r 9.1.1) that faithfully reflects preferences relative to a preexisting significance axis. Such a preference model is called an expected utility criterion.

---

[8] *translator's note: A further example is provided on Page 239 of the original, French version at this point.*

To construct such a utility function, one could proceed as follows. Determine the degrees e, and for each, assign a probability π that leads to indifference between the two following distributions:

- a degenerate distribution located at degree e;
- a distribution consisting of only the highest degree of E_i (denoted e^*) with probability π and the lowest degree of E_i (denoted e_*) with probability $1 - \pi$.

By convention, one can assume:

$$v(e^*) = 1 \text{ and } v(e_*) = 0.$$

Applying this preference model to the two distributions, then, leads to:

$$v(e) = \pi \cdot v(e^*) + (1 - \pi) \cdot v(e_*) = \pi,$$

which uniquely defines the utility v(e).[9]

It is interesting to note that discounting theory also tries to overcome the same types of difficulties. The modulation index is not a probability in this case, but can be defined by modeling the way in which a sum S(a) is dispersed along a time axis. Under these conditions, the present value of the sum using a discount rate j is:

$$\sum_{e \in \gamma_i(a)} \frac{\delta_i^a(e)}{(1 + j)^e},$$

$$\text{with} \sum_{e \in \gamma_i(a)} \delta_i^a(e) = S(a).$$

Once again, we have a point reduction criterion of the type (r 9.2.5) where $v(e) = \dfrac{1}{(1 + j)^e}$.

Just as in the probabilistic case, it is easy to show that the function v(e) results from situations of indifference along the significance axis: The quantity S in period e is considered equivalent to the quantity $\dfrac{S}{(1 + j)^e}$ in period 0.

---

[9] *We discuss this further in Section 9.4.4.1; interested readers should be able to follow the discussion there at this point.*

b) Point reduction based on percentiles or on other dispersion characteristics[10]

Either because the requirements stated at the beginning of a2) are not satisfied or for some other reason, one might need to base the point reduction criterion on other important characteristics of the dispersion. One family of such characteristics consists of quantities called **percentiles** in statistics.

Recall that the percentile p (p being a whole number between 0 and 100) is the degree $e^p(a)$ such that:

$$100 \times \frac{\sum_{e \leq e_p(a)} \delta_i^a(e)}{\sum_{e \in \gamma_i(a)} \delta_i^a(e)} \text{ is as close as possible to p.} \qquad \text{(r 9.2.6)}$$

(Since the scale is discrete, there may not be a degree e_p that allows exact equality with p; this difficulty would generally not arise with continuous scales.) If δ is not additive (see Section 8.2.3.5), e_p would likely have no real meaning. Finally, if e_p is to be a point reduction criterion, the denominator of the above expression must (with minor exceptions) be independent of a: otherwise it would be rare that e_p could correspond to a totalitarian significance axis.

In practice, the percentiles most often used as point reduction criteria are:

- the median $e_{50}(a)$, which often has certain advantages over the average;
- the first quartile $e_{25}(a)$ and the third quartile $e_{75}(a)$;
- the first decile $e_{10}(a)$ and the ninth decile $e_{90}(a)$, which are often used when the impact of the elementary consequence c_i on preferences is mostly due to very low or very high degrees of the state indicator.

When the dispersion index is not additive (see Sections 8.2.3.1 and 8.2.3.2), the point reduction must be based on more summary characteristics of the dispersion. The best known are:

$$e_{min}(a) = \min\{e/e \in \gamma_i(a)\},$$
$$e_{max}(a) = \max\{e/e \in \gamma_i(a)\},$$
$$e_{mode}(a) = \max_{e \in \gamma_i(a)} \delta_i^a(e) \text{ (degree which is not necessarily unique).} \qquad \text{(r 9.2.7)}$$

Important information is often lost when using one of these degrees as a point reduction criterion. We shall see in Section 9.3 that introducing discrimination thresholds can be a means of reducing this loss of information.

---

[10] *We shall assume that $E_i \subset \mathbb{R}$, although other encodings could be possible.*

Depending on the specific characteristics of the problem, many other solutions could be considered. An illustration is provided in the following example.[11]

Example 2: Commuter Rail Line (from Section 8.2.4)

Consider first the unreferenced modulation indicator related to trip time savings defined in Section 8.2.3.1. This defines a point criterion with a straightforward interpretation:

$$g(a) = \frac{\gamma^-(a)\cdot\delta^a(\gamma^-(a)) + \gamma^0(a)\cdot\delta^a(\gamma^0(a)) + \gamma^+(a)\cdot\delta^a(\gamma^+(a))}{\delta^a(\gamma^-(a)) + \delta^a(\gamma^0(a)) + \delta^a(\gamma^+(a))}.$$

This point reduction is based on a procedure that combines the three representative degrees defined in (r 9.2.8), $e_{min}(\gamma^-(a))$, $e_{mode}(\gamma^0(a))$, and $e_{max}(\gamma^+(a))$. The extreme degrees e_{min} and e_{max} would normally be multiplied by small weights in order to allow the modal degree to have a dominant role in the criterion.

Consider now the scenarios discussed in Section 8.2.4. The point reduction procedure presented above could be applied to each scenario s, leading to a number g(a, s). Depending on the circumstances, the analyst could then:

- perform the point reduction by considering the worst scenario:

$$g(a) = \min_{s \in S} g(a, s);$$

- perform the point reduction by considering weights or subjective probabilities p(s) for each scenario and calculating an average:

$$g(a) = \sum_{s \in S} p(s)\cdot g(a, s);$$

- not attempt to perform any further point reduction: g(a, s) values could form a criterion, leading to one criterion for each scenario; in this case, there would be both point reduction on and splitting of the dimension.

(end of Example 2)[12]

c) Point equivalent: another look at utility theory

Again, assume that $E_i \subset \mathbb{R}$, and denote e_* as the lowest degree of E_i and e^* as the highest degree. To present the concept of point equivalent, we have to consider any number e that satisfies $e_* \leq e \leq e^*$ as a degree of E_i. This may require adding degrees to E_i – for example, if the scale is discrete. These serve only as artificial degrees for interpolation, however, with no role in evaluating actions. As before we shall limit ourselves to the case where $\gamma_i(a)$ is a finite subset of E_i. This limitation is not very restrictive in practice, and the increased rigor that is obtained by relaxing it complicates the presentation with no real practical merit.

---

[11] *translator's note: The continuation of Example 9 on page 243 of the original, French version is provided as an additional illustration.*

[12] *translator's note: Example 2 is continued and ended in Section 12.3 in the original, French version.*

Consider an operator G that is used to define a criterion g:

$$g(a) = G[\gamma_i(a), \delta_i^a(e)],$$

(see (r 9.2.4)). Let $C_i(g, a)$ be the set of couples (γ_i, δ_i) for which G is defined and such that:

$$G[\gamma_i, \delta_i] = g(a).$$

For point reduction operators G normally used to define a point reduction criterion for any action a considered, the equation:

$$G[\gamma_i(a), \delta_i^a(e)] = G[\varepsilon, \delta^\varepsilon] \qquad \text{(r 9.2.8)}$$

has at most one solution in ε, where, δ^ε denotes the degenerate distribution with the quantity $\sum_{e \in E_i} \delta_i^a(e)$ at degree ε.[13]

When it exists, this solution, which is an element of $C_i(g, a)$, is denoted $\varepsilon_g(a)$. The degree $\varepsilon_g(a)$ is, by definition, the **point equivalent of action a on criterion g**. This concept is a generalization of that of certainty equivalent introduced in utility theory (see footnote 2 below.) The term point equivalent represents the fact that, for the criterion g, everything is as if action a only had $(\varepsilon_g(a), \delta^\varepsilon)$ for its evaluation.

It is possible that $\varepsilon_g(a) = g(a)$ for all a. Specifically, this is so when G implies calculating the average of the degrees weighted by their dispersion index values (see, for example, the beginning of a1)). This might lead one to think that $\varepsilon_g(a)$ could be chosen as a criterion function in other cases as well.[14]

9.2.2.2 Splitting dimension i

Every point reduction criterion leads to a single value for the various $[\gamma_i, \delta_i]$ evaluations. Because of the complexity of the consequence c_i, however, these might be internalized differently by the different actors. In these cases, one may wish to split the dimension i to avoid certain actors' rejecting the point reduction criterion or to postpone introducing subtleties implicitly contained in the point reduction. Less of the information contained in the evaluation is lost when performing the point reduction in this way. Also, it avoids covering up the comparison difficulties associated with the dispersion. On the other hand, the number of criteria increases, and this can cause certain problems, as we shall see later.

---

[13] *translator's note: The motivation for this stratement is briefly discussed in the original, French version on pp. 244-245.*

[14] *translator's note: Two examples, which also illustrate the concept of a point equivalent, are presented on Pages 245-247 or the original, French version to address this point.*

We conclude this subsection by illustrating how some of the dimensions already discussed can be split.

First, consider a discomfort scale P (Section 8.1.3.2) associated with mass transportation. Assume that this scale has four degrees: e_0, e_1, e_2, and e_3, where e_0 represents a state that is considered uncomfortable; e_1 represents a state that is not uncomfortable, but does not permit any activity; e_2 represents a state that allows certain activities (e.g., reading a newspaper, easy knitting, light conversation, ...) to be conducted, but at the price of some fatigue; and e_3 represents a comfortable state that would allow more complicated activities (e.g., reading complex material, difficult knitting, professional conversation, ...) to be conducted. Let $\delta_i^a(e_h)$ be the number of people who will be placed at degree e_h if a is implemented (e.g., number of people transported with conditions e_h). In splitting this dimension, one might consider the two following criteria:

$$g(a) = \delta_i^a(e_0) + \delta_i^a(e_1),\ g'(a) = \delta_i^a(e_3).$$

These criteria, which do not require any encoding of the scale P, are easily understood: number of people transported in relatively poor conditions, number of persons transported in very comfortable conditions. Note that, even though they are not completely independent, the criteria have no redundancy, even if $\sum_h \delta_i^a(e_h)$ is independent of a.

Let us now reconsider receiving a monetary sum S(a) distributed in several payments. The amount of the payment in the period e is $\delta_i^a(e)$. Let g(a) be the present value at rate j of these payments. Series of equal, increasing, or decreasing payments, with little or much dispersion in time, can all result in the same value for this criterion. The individual receiving the payments might feel, however, that payments to be received in the distant future are associated with more risk (loss of solvency of the source of payment, inflation, ...). Therefore, when comparing two actions a and a′ with $g(a) = g(a')$, she might have a distinct preference for the one that pays out the amount sooner. If the series of payments associated with a is much more dispersed in time than that associated with a′, she might prefer a′ to a even if $g(a') < g(a)$. This would imply that the significance axis of criterion g is not a totalitarian axis (Section 9.1.2).

To capture the preferences associated with a dimension correctly under these conditions, it is useful to complement the significance axis of g with another axis reflecting the dispersion of payments in time. The present value criterion is, then, only one splitting criterion of the dimension, where others could be defined in several ways, for example:

$g'(a)$ = period in which the sum of the payments reaches a given fraction (e.g., 90 percent) of S(a);

$g'(a)$ = period in which the present value at rate j of the payments reaches a given fraction (e.g., 95 percent) of g(a).

Problems dealing with investment choices usually involve a period of net expenses followed by a period of net revenues. This fact and the above considerations explain the numerous criteria (see, e.g., Holl, *et al.*, 1973) that have been proposed to compare investments. Some authors or practitioners discuss the respective merits of these criteria in an attempt to arrive at a "best criterion." In many cases, however, several of these criteria can be used jointly to split a financial dimension. The issue becomes even more complicated when one considers different scenarios that can cause variations in the series of payments (referenced dispersion indicators; see Rizzi, 1984).

Let us finally consider a dimension i used to denote time savings obtained by a certain population as a result of possible actions. Let E_i be a discrete scale (e = number of minutes saved). One might wish to use a dispersion index $\delta_i^a(e)$ that reflects the number of people who would have a time savings e if action a were executed. In many applications (comparing schedules or operational policies; see also Example 8), it is difficult to accept a comparison of two actions based on a totalitarian significance axis that corresponds to a point reduction, such as the average or median time saved. Such a criterion could assign equal values to two actions even though one of them might lead to a large fraction of the population's having no, or even negative, time savings while the other leads to positive time savings for everyone. One simple way to handle this aspect in comparisons is to split the dimension by using percentiles. For example, one might use e_{10} and e_{50}; or e_{10} and e_{90}; or e_{10}, e_{50}, and e_{90}, defined above (r 9.2.7).

9.2.3 Criterion function based on a subset of dimensions

In constructing a criterion function that encompasses a subset I of the dimensions (which we denote g_I), there are two distinct cases:

- the evaluation is a point evaluation on each of the dimensions of I;
- the evaluation is a nonpoint evaluation on at least one of the dimensions of I.

In the first of these two cases, criterion g_I aggregates the point state indicators γ_i over the I dimensions. Each of these state indicators could be considered a criterion associated with dimension i, and g_I could thus be considered the aggregation of the criteria g_i related to the individual dimensions. The same would be true in the second case if the construction of g_I can be decomposed into two phases: a first phase, where a point reduction criterion is introduced for each of the dimensions in which there are nonpoint evaluations; followed by a second phase, which operates only on these criteria and on the state indicators of those dimensions with point evaluations.

These observations illustrate the relationship that can exist between the conception of a unique criterion aggregating several separable and previously constructed criteria and the conception of a criterion g_I directly formed from state and dispersion indicators on the dimensions of I. This direct conception can often offer distinct advantages, such as simplified procedures, reductions in the number of criteria, and clearer communication. To put it into practice, however, it is important to verify that the following conditions hold.

CONDITION I: The dimensions of I are sufficiently similar or complementary in their elementary consequences so that the significance axis g_I is transparent to and easily understood by the different actors.

CONDITION II: None of the actors disagrees with the synthesis of the different dimensions brought about by g_I. (Possible disagreements might arise either from the manner in which the synthesis is performed or from the values of the parameters used, and would probably indicate a difficulty with the value system underlying the synthesis.)

There may be times, of course, when the analyst will wish to use a criterion whose support encompasses some dimensions that do not fulfill these two conditions. In this case, it is usually preferable to think of such a criterion as aggregating several criteria, each of which is linked to a single dimension of I or to a subset of the dimensions of I. We shall thus speak of **criteria aggregation**. Although we do not discuss the difficulties that can arise from this aggregation until Chapter 11, we mention here that these difficulties are essentially the same whether all or only a few criteria violate one or both of the conditions presented above. On the other hand, the difficulties are drastically reduced when the conditions are satisfied.

We propose the following four categories of circumstances that frequently justify the direct conception of a criterion of the type g_I.

a) One dimension is dominant among the I dimensions

In this first category of circumstances, the dimensions of I other than some dominant dimension i_0 concern consequences that play a secondary, "corrective" role to the value taken on by the criterion that would result from evaluating only along dimension i_0.

Suppose, for example, that I consists of two dimensions. One dimension gives rise to a point evaluation γ_1 on a scale E_1 that is formed by real numbers between 0 and 100, such as a ratio in the loan application example. The other dimension is designed to capture some qualitative information that amplifies or diminishes the impact of the evaluation on γ_1. The scale might be $E_2 = \{++, +, 0, -, --\}$. The analyst might then interpret the impact of the evaluation on the second dimension by proposing:

$$g_I(a) = \begin{cases} \gamma_1(a), \text{ if } \gamma_2(a) = 0 \\ \gamma_1(a) + 2, \text{ if } \gamma_2(a) = + \\ \gamma_1(a) - 1, \text{ if } \gamma_2(a) = - \\ \gamma_1(a) + 2 + \dfrac{100 - \gamma_1(a)}{20}, \text{ if } \gamma_2(a) = ++ \\ \\ \gamma_1(a) - 1 - \dfrac{\gamma_1(a)}{20}, \text{ if } \gamma_2(a) = -- \end{cases}$$

The asymmetries in the above relation would, of course, be the result of specific phenomena that represent a consensus among the actors.

The lexicographic technique for constructing a criterion g_I would be another example related to this first category. This technique concerns the case where the elementary consequences c_i for $i \neq i_0$ are only considered when the different actions are identical according to c_{i_0}.

b) I consists of two or three dimensions whose scales can be reduced to a few degrees

When I consists of only a few (two or three) dimensions whose scales can be reduced to a few degrees, the analyst may try to define g_I by first constructing a criterion g_i with values on the scale E_i for each of the dimensions. (For the point indicator case, $g_i = \gamma_i$.) Then, he could consider the elements of E_I to be the couples or triplets that are formed by the Cartesian product of the E_i scales. He could then consider the preferences of the actors (see Condition II) to form a complete preorder on E_I that would define g_I.

This would be as if the original dimensions play the role of subdimensions with respect to the new dimension identified by the significance axis g_I. More generally, one could consider the "méthodes des déclassements comparés" (Le Boulanger and Roy, 1970.)

c) The elementary consequences associated with I are evaluated on the same scale E_I

As an example, consider a series of grades on a scale of 0 to 10 given to a student in different subjects (see, e.g., Example 10). It is easy to transform these grades into a single synthesized grade by calculating the weighted average in each subject (which would be a point reduction on each of the dimensions) and then a weighted average of these averages. This common practice is representative of the category of circumstances considered here and assumes that those responsible for judging students accept both the compensatory mechanism of the average (a point to which we return in Chapter 11) and those elements in the value system that form the basis of the different weighting schemes – namely, weightings according to the types of grades in each subject and weightings of the subjects.

Another traditional example concerns elementary consequences, each of which gives rise to a monetary scale. The criterion g_I could be interpreted as a total cost that integrates the different categories of expenses. As in the preceding example, the reader can see that the presence of a common scale E_I is pivotal to the definition g_I. It does not solve all of the problems, however. Simply because investment and operating costs are expressed in the same units does not mean that adding the two together will define a criterion that adequately synthesizes these two elementary consequences. The costs would have to cover the same period. They would also have to be such that they do not represent different costs to different actors – e.g., the government and taxpayers; producers and consumers.

d) The elementary consequences associated with I lead to a natural synthesis for reasons other than those given in a), b), c)

In the context of media-planning (Example 5), for example, it is common to refer to the "cost per thousand targeted readers." This could form a significance axis on which one could define a criterion on the basis of a quotient of only two elementary consequences: the price of a standard advertisement and the targeted audience. The number defined in this way may not be pertinent, however, since the different periodicals can all behave differently on two other dimensions: the likelihood of being seen and the attention they generate. Instead of dividing the cost of the standard advertisement by the targeted audience, it may, therefore, be preferable to divide the cost by the number of people in this targeted audience who had the chance to see the advertisement at least one time after it appeared some number m times in the periodical. Such a criterion would, therefore, use four dimensions.

What is often called a generalized cost or a generalized time, especially in the field or passenger transportation, could fall in this fourth category of circumstances. We would point out that such a criterion is only acceptable when the value system that allows a combination of real costs or times with certain characteristics of the trip (transfers, noise, ...) is not open to too much debate. This emphasizes once again the need to fulfill Condition II. Other examples can be found in Roy *et al.* (1986) and Roy and Slowinski (1993).

To conclude this section we emphasize the following points dealing primarily with terminology.

Using a criterion of the type g_I does not exclude the use of some of the dimensions of I in the support of other criteria. Thus in the Media Planning example (see section 8.1.5, Example 5), in addition to the cost of reaching 1000 targeted readers, there is also a criterion reflecting the strength of the periodical that is based on the target population reached by the periodicals. When the dimensions of I are only used in the support of g_I – i.e., when they are not used in any other criterion – we say that g_I is a **sub-aggregating criterion of the dimensions of I**.

As shown in the preceding examples, Condition II can be considered to be fulfilled even if the means of forming the state and dispersion indicators that help define g_I is somewhat arbitrary. Nevertheless, this arbitrariness must be sufficiently limited so that it can be handled by discrimination thresholds, a concept that we introduce in the next section.

Even more than in the cases of the criteria discussed in Sections 9.2.1 and 9.2.2, it is particularly important that the significance axis of a criterion g_I be understood easily and without ambiguity (Condition I). Indeed, in the preceding cases there was only a single elementary consequence, the significance axis was closely linked to the dimension i, and the criterion usually took on values on a scale E_i or on a set resulting from interpolation or encoding. In the cases discussed in this subsection, there are several dimensions and usually several scales.

Whether criterion g is based on one or several dimensions, its possible values define a completely ordered set that reflects preference associated with the significance axis of criterion g (see Def. 9.1.1, remark b). Indeed, when considering two ideal actions a and a′, respectively, for which the comparison corresponds exactly to two values e and e′ (i.e., no approximation, inaccurate determination, or arbitrariness that could affect the values of the criteria), it is not restrictive to propose that:

$$a \; I_g \; a' \text{ if } e = e';$$
$$a \succ_g a' \text{ if } e > e'.$$

This results from the fact that Z accepts the validity of the model comprised of the criterion g that allows the comparison of two actions along the significance axis used to define g (see Def. 9.1.1, remark b).

Thus, the set of possible values for a criterion g can always define a scale in the sense of Definition 8.1.3. We shall, therefore, denote this scale $\mathbf{E_g}$ and speak of the **scale associated with criterion g**, or more simply, the **scale of g**.

Finally, notice that, when g is based on a set I of dimensions, E_g may very well be different from each of the E_I scales.

9.3 TRUE CRITERIA, SEMI-CRITERIA, PRE-CRITERIA, PSEUDO-CRITERIA

We now return to the two specific points in the definition of a criterion function g. From Definition 9.1.1, this function must be such that:

$$g(a') \geq g(a) \Rightarrow a' \ S_g \ a.$$

That is, g reflects outranking situations. These situations cover situations of indifference, weak preference, and strict preference. It is also possible that certain outranking relations are not reflected by the values g(a′) and g(a): $g(a') \geq g(a)$ does not exclude $a \ S_g \ a'$, since the implication is only to the right. It follows that $g(a') \neq g(a)$ is not necessarily incompatible with the assertion $a' \ I_g \ a$. We shall now try to capture the behavior of g relative to these two possibilities (or the hypotheses that one formulates relative to this behavior) through the notion of the discriminating power of a criterion function.

First of all, we specify this notion based on the concept of a true criterion, which corresponds to the most traditional behavior modeled by a criterion function. This will lead us to distinguish in Section 9.3.3 three other forms of criteria with slightly different discriminating power. The last subsection uses examples to illustrate means of determining discrimination thresholds.

9.3.1 True criterion and discriminating power of a criterion

*DEFINITION 9.3.1: A **true criterion** is a criterion-function g such that:*

$$g(a') \geq g(a) \Rightarrow \begin{cases} a' \ I_g \ a \text{ if } g(a') = g(a); \\ a' \ P_g \ a \text{ if } g(a') > g(a). \end{cases}$$

This definition implies that for a true criterion:

- there is indifference between a and a′ along the significance axis of g only when $g(a) = g(a')$;
- every positive difference $g(a') - g(a)$ corresponds to a strict preference along the significance axis of g for a′.

Therefore, a function g is a true criterion if and only if it satisfies the two implications of (r 9.1.1) that were presented in Section 9.1.2 as the most traditional way of defining a criterion. Whether it be in engineering economics, management science, or operations research, the concept of criterion has long been taken to be that of a true criterion, a criterion that acknowledges only situations of indifference and strict preference, and rejects situations of weak preference. Therefore, we shall say that a true criterion possesses **absolute discriminating power**.

There are many reasons that the analyst might not wish to consider a small difference $g(a') - g(a) \neq 0$ as not signifying strict preference, however. Among them, we mention that : i) the values given to state and dispersion indicators forming the support of g will usually be affected by imprecision, uncertainty, and inaccurate determination; ii) the calculation process that determines these values may be questionable; iii) even with ideal actions, where the other problems are assumed not to exist, the decision maker may consider slight variations in parameter values and other assumptions to be insignificant and, therefore, feel that nonzero differences in $g(a') - g(a)$ are not always sufficient to elicit strict preference. Under such conditions, if the analyst wishes to refine the meaning of the relations S_g in the implication:

$$g(a') > g(a) \Rightarrow a' \; S_g \; a,$$

he must admit that:

- P_g can only replace S_g when the difference $g(a') - g(a)$ is sufficiently large;
- I_g or Q_g can replace S_g when the difference $g(a') - g(a)$ is sufficiently small.

The analyst is thus led to consider forms of criteria that are no longer true criteria and that we shall say possess a **nonabsolute or a nuanced discriminating power**.

The discriminating power of criterion function g is related to how the analyst sees the ability of g to distinguish among situations of strict preference, indifference, and weak preference on the basis of the magnitude of the difference $g(a') - g(a)$.

It is important to note that, for a given criterion g, the analyst has options as to how to discriminate among the situations I_g, Q_g, P_g (or combinations of them) covered by S_g (see Section 4.2). Thus, he always has the freedom to consider g as a true criterion, but he must then wonder whether the basic system of preference relations (I_g, P_g) associated with it would really reflect the preferences over A that are being modeled. The entire decision aiding approach, and consequently, the persuasiveness of its conclusions, are influenced by the options taken at this level. Blindly accepting the absolute discriminating power of g could, for example, lead to inappropriate overconfidence in the results.

When the analyst feels the necessity to choose a more nuanced discriminating power, he can simply do so by introducing one or two thresholds, as we show next. We mention now, however, that these thresholds, called discrimination thresholds of g, are similar to dispersion thresholds of a state indicator in certain ways, but are different in other ways. The simplicity of the resulting preference models is based on a conception of discriminating power that is not completely general, since it allows only one factor to influence the importance of the size of the difference g(a′) − g(a), namely, the position of the interval [g(a), g(a′)] on the scale E_g. To understand this restriction better and to illustrate the notion of the discriminating power of a criterion, we use the rest of this subsection to discuss the important case where g is a point reduction criterion along dimension i.

Assume the following conditions:

- $\gamma_i(a)$ is an interval of E_i, and $\delta_i^a(e)$ represents the importance, likelihood, or probability, of each degree e of $\gamma_i(a)$ (consider, for example, "future costs/current costs" in Section 8.2.2.2, Example 6; "installation date" in Section 8.2.3.1, Example 4);
- g is a point reduction criterion that represents a central value of δ_i, such as a mode, median, mean, point equivalent, ... (see Section 9.2.2), and, therefore, satisfies $g(a) \in \gamma_i(a)$.

To determine the cases where S_g corresponds to P_g and I_g, it makes sense to accept:

- $g(a') - g(a) > 0$ represents a situation of strict preference when the intervals $\gamma_i(a')$ and $\gamma_i(a)$ are disjoint or when their intersection is an interval where the size or the likelihood seems "negligible" (see Fig. 9.3.1);
- $g(a') - g(a) \geq 0$ represents a situation of indifference when $g(a') \in \gamma_i(a)$ and $g(a) \in \gamma_i(a')$, and the positions of g(a′) and g(a) in each of these intervals are sufficiently central so that the sizes or the likelihoods of the intervals that separate them from the extremities of these intervals are "nonnegligible" (see Fig. 9.3.2).

Figure 9.3.1: Example of strict preference with a point reduction criterion g on dimension i

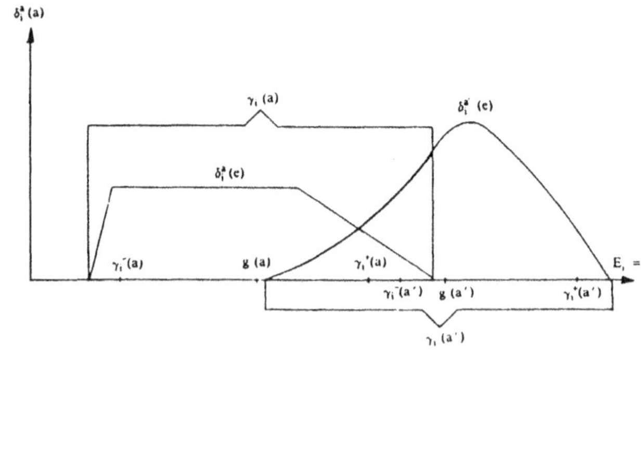

To provide a more precise description of the cases $S_g = P_g$ and $S_g = I_g$, we must specify what we mean by an interval of negligible size or likelihood. To do so, we could use a simple rule to cut off the largest possible interval at each of $\gamma_i(a)$'s extremities that is

considered negligible according to $\delta_i^a(e)$. This truncation leads to a new interval $[\gamma_i^-(a), \gamma_i^+(a)]$ in $\gamma_i(a)$ (see Figs. 9.3.1 and 9.3.2). We can then write:

$$g(a') - g(a) > 0 \text{ and } [\gamma_i^-(a'), \gamma_i^+(a')] \cap [\gamma_i^-(a), \gamma_i^+(a)] = \varnothing \Rightarrow a' \, P_g \, a; \qquad \text{(r 9.3.1)}$$
$$g(a') - g(a) \geq 0, \; g(a') \in [\gamma_i^-(a), \gamma_i^+(a)] \text{ and } g(a) \in [\gamma_i^-(a'), \gamma_i^+(a')] \Rightarrow a' \, I_g \, a. \qquad \text{(r 9.3.2)}$$

Figure 9.3.2: Example of indifference with a point reduction criterion g on dimension i

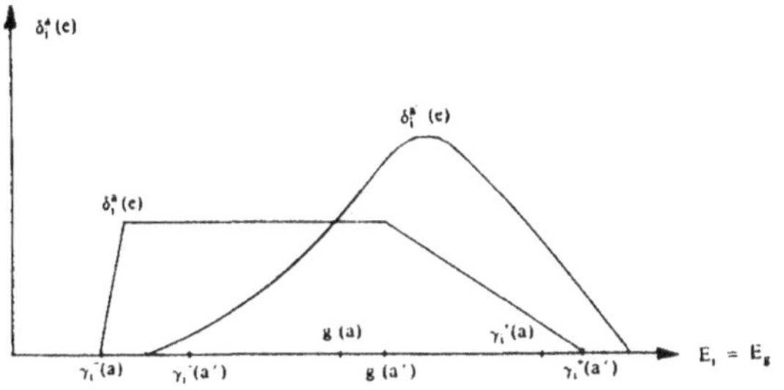

Let:

$$g(a) - \gamma_i^-(a) = \eta_i^-(a) \text{ and } \gamma_i^+(a) - g(a) = \eta_i^+(a).$$

Using Figures 9.3.1 and 9.3.2, the reader can see that (r 9.3.1) and (r 9.3.2), respectively, are equivalent to:

$$g(a') - g(a) > \eta_i^+(a) + \eta_i^-(a') \Rightarrow^{15} a' \, P_g \, a; \qquad \text{(r 9.3.3)}$$
$$0 \leq g(a') - g(a) \leq \text{Min}\{\eta_i^+(a), \eta_i^-(a')\} \Rightarrow a' \, I_g \, a. \qquad \text{(r 9.3.4)}$$

(One can obtain the relations concerning g(a) > g(a') by switching the order of a and a'.) To finish defining the discriminating power of g, we let:

$$\text{Min}\{\eta_i^+(a), \eta_i^-(a')\} < g(a') - g(a) \leq \eta_i^+(a) + \eta_i^-(a') \Rightarrow^{1} a' \, Q_g \, a. \qquad \text{(r 9.3.5)}$$

Relations (r 9.3.3), (r 9.3.4), and (r 9.3.5) form a relatively general model that provides a specific interpretation to the concept of discriminating power of a criterion. The quantities $\eta_i^+(a) + \eta_i^-(a')$ and $\text{Min}\{\eta_i^+(a), \eta_i^-(a')\}$, which must be compared to the difference g(a') − g(a) to determine situations of strict preference and strict indifference, depend on $\delta_i^a(e)$ and $\delta_i^{a'}(e)$ in a complex way. That is, without any other hypothesis,

[15] *In this model, the implication can be replaced by "if and only if" without any loss of generality.*

these quantities cannot be defined only as a function of the position of the interval [g(a), g(a′)] on E_g.

9.3.2 Indifference and preference thresholds

Let g be a criterion, and let a and a′, respectively, be a fixed and variable action in A. Consider a′ such that g(a′) – g(a) = 0. By definition this implies a′ I_g a. Next, consider a′ to vary such that g(a′) increases. The analyst might accept a′ I_g a for insignificant, but nonzero differences g(a′) – g(a). However, when the difference becomes large enough, he will no longer accept this relation. Assume that the analyst can either use theoretical considerations or common sense to fix an upper limit q_g on the value of the difference g(a′) – g(a) such that any difference less than q_g is not sufficient to opt for a preference (not even a weak preference) in favor of a′.

Applying this line of reasoning to the case presented at the end of Section 9.3.1, we notice that the limit q_g might depend on g(a), but that it could also depend, through the dispersion indices, on other factors associated with a or on the characteristics of the various actions a′ considered (see Fig. 9.3.2). In this subsection, we will consider the former case – that is, where the limit q_g is a function $q_g[g(a)]$ intrinsically linked to the scale E_g. This function is, therefore, called an **indifference threshold associated with criterion g**. This threshold could be independent of g(a), or it could increase (e.g., proportionally) with it. It might also decrease with g(a), but we shall later specify certain restrictions for this case.

Based on the above, we can define the indifference threshold q_g by stating:

$$\forall\ a, a' \in A,\ 0 \leq g(a') - g(a) \leq q_g[g(a)] \Rightarrow a'\ I_g\ a, \text{ and}$$
$$g(a') - g(a) > q_g[g(a)] \Rightarrow a' \succ_g a, \qquad \text{(r 9.3.6)}$$

where $\succ_g$ is the preference relation (see Table 7.1.5) along the significance axis of g.

Assume now that g(a) is far enough from the upper extremity of the scale of criterion g that there could be an action (perhaps a dummy action) a′ such that a′ P_g a. Let us now vary a′ in such a way that g(a′) decreases. At some point, the difference g(a′) – g(a) will become too small to remain compatible with a′ P_g a. Let us assume that the analyst can fix a lower limit p_g on the value of the difference g(a′) – g(a) such that any difference less than this limit is not sufficient to opt for a strict preference in favor of a.

As with q_g, this new limit p_g might depend on a through other factors of g(a) and might also depend on a′ (see Fig. 9.3.1). In this subsection, we will consider the case where the limit p_g is intrinsically linked to the scale E_g. The function $p_g[g(a)]$ is, therefore, called a **preference threshold associated with criterion g**. Again, this threshold may be independent of g(a), or it could increase or decrease (with restrictions that we shall specify later) as a function of g(a).

Based on the above, we can define the preference threshold p_g by stating:

$$\forall\, a, a' \in A,\ g(a') - g(a) > p_g[g(a)] \Rightarrow^{16} a'\ P_g\ a, \text{ and}$$
$$0 \le g(a') - g(a) \le p_g[g(a)] \Rightarrow a'\ J_g\ a, \qquad \text{(r 9.3.7)}$$

where J_g is the J-preference relation of Table 7.1.5.

From the definitions of q_g and p_g, it follows that:

$$\forall\, a \in A,\ 0 \le q_g[g(a)] \le p_g[g(a)].$$

From these two definitional relations of thresholds, we can easily deduce:

$$\forall\, a \in A,\ q_g[g(a)] < g(a') - g(a) \le p_g[g(a)] \Leftrightarrow a'\ Q_g\ a.$$

(The reader can notice the conventions that are assumed to determine the situations when $g(a') - g(a)$ is equal to one of the thresholds.) The diagram in Figure 9.3.3 summarizes the preceding results.

Figure 9.3.3: Indifference, weak difference, and strict preference zones on E_g when $g(a') \ge g(a)$

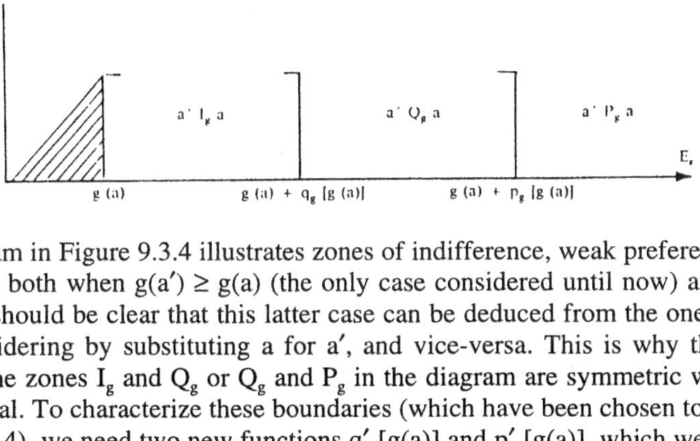

The diagram in Figure 9.3.4 illustrates zones of indifference, weak preference, and strict preference both when $g(a') \ge g(a)$ (the only case considered until now) and when $g(a') < g(a)$. It should be clear that this latter case can be deduced from the one that we have been considering by substituting a for a′, and vice-versa. This is why the boundaries between the zones I_g and Q_g or Q_g and P_g in the diagram are symmetric with respect to the diagonal. To characterize these boundaries (which have been chosen to be rectilinear in Fig. 9.3.4), we need two new functions $q'_g[g(a)]$ and $p'_g[g(a)]$, which we explain next.

The functions q_g and p_g were defined with the less preferred value of g(a) and g(a′) as the argument. Let q'_g and p'_g represent the functions one would obtain if the thresholds had been expressed as a function of the more preferred of these two values.

To see how these two new functions can be derived from q_g and p_g, we still consider g(a) constant and decrease g(a′) from a level g(a) in a way such that it satisfies:

$$g(a') + q_g[g(a')] \ge g(a).$$

[16] *Again, this implication can be replaced by "if and only if".*

In practice E_g, whether a discrete or continuous scale, is always bounded and there is, therefore, a smallest value of g(a′) that satisfies this condition. Call this value y[g(a)] and assume that the required condition is never violated when decreasing g(a′) from g(a) to y[g(a)], a reasonable hypothesis to which we shall return (see (r 9.3.10)). It follows that every action a′ for which g(a′) = g(a) – q′ (q′ ≥ 0) is compatible with a′ I_g a, as long as q′ ≤ g(a) – y[g(a)]. Therefore, we have:

$$q'_g[g(a)] = g(a) - y[g(a)]. \qquad \text{(r 9.3.8)}$$

Figure 9.3.4: Indifference, weak preference, and strict preference zones in the g(a), g(a′) plane when E_g = [0, 1]

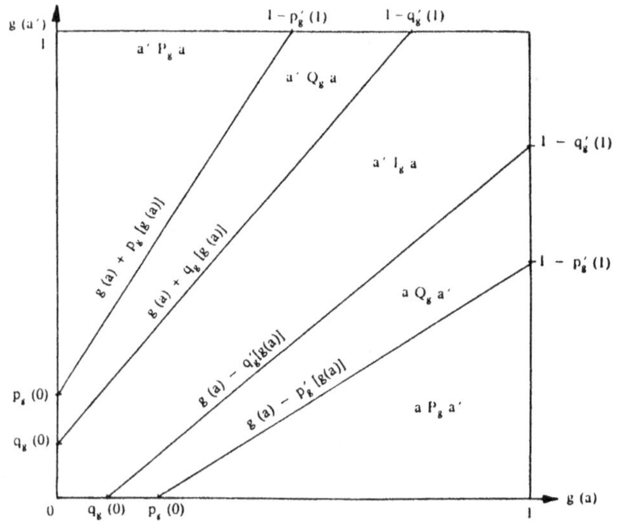

In the same way, let a function z[g(a)] give the smallest value of g(a′) satisfying:

$$g(a') + p_g[g(a')] \geq g(a).$$

This leads to the second function:

$$p'_g[g(a)] = g(a) - z[g(a)]. \qquad \text{(r 9.3.9)}$$

We shall call the functions q'_g and p'_g, respectively, **the inverse indifference threshold** and **the inverse preference threshold associated with g**. To avoid ambiguity, we could qualify the thresholds q_g and p_g as being **direct** thresholds to emphasize that they were established on the basis of increasing preference. **We note that direct thresholds are only equal to inverse thresholds when the thresholds are constant** (in absolute value). In the case where the thresholds are proportional to g(a), the reader can verify that:

$$\text{if } q_g[g(a)] = q \cdot g(a) \ (q \geq 0),$$
$$\text{then } y[g(a)] = \frac{g(a)}{1 + q},$$

$$\text{and } q'_g[g(a)] = \frac{q}{1 + q} g(a).$$

Let us now see what the preceding definitions imply for increases or decreases in the functions $q_g[g(a)]$ and $p_g[g(a)]$. Consider two actions a and a′ such that a′ P_g a. Let b be an action such that $g(b) \leq g(a)$. It would, therefore, be reasonable to assume that a′ P_g b, which is equivalent to:

$$g(b) \leq g(a) \Rightarrow g(b) + p_g[g(b)] \leq g(a) + p_g[g(a)].$$

(If this implication did not hold, there could exist g(a′) compatible with a′ P_g a and not a′ Q_g b, which is not permitted.) Note that this implication is also equivalent to requiring that $g(a) + p_g[g(a)]$ **be a nondecreasing monotonic function of g(a)**.

Similarly, we can arrive at:

$$g(b) \leq g(a) \Rightarrow g(b) + q_g[g(b)] \leq g(a) + q_g[g(a)],$$

which requires that $g(a) + q_g[g(a)]$ be a nonincreasing monotonic function of g(a), the hypothesis used in establishing (r 9.3.8). The reader can easily show that these conditions of monotonicity are equivalent to:

$$\frac{q[g(a)] - q[g(b)]}{g(a) - g(b)} \geq -1, \quad \frac{p[g(a)] - p[g(b)]}{g(a) - g(b)} \geq -1. \qquad \text{(r 9.3.10)}$$

Although the four boundaries of the five zones in Figure 9.3.4 do not have to be rectilinear, the considerations developed above impose certain restrictions: The functions $q_g[g(a)]$ and $p_g[g(a)]$ can increase without restriction, but they must not decrease too rapidly.

Relation (r 9.3.10), with the reasoning that led to it, highlights certain analogies that exist between indifference or preference thresholds – which we shall call discrimination thresholds, in general – and intrinsic dispersion thresholds defined in Section 8.2.2.3. On the other hand, the functional relation that exists between direct and inverse thresholds in the case of **discrimination thresholds** does not necessarily exist for positive and negative dispersion thresholds (see Section 8.2.2.1).

9.3.3 Pseudo-criteria, semi-criteria, pre-criteria

Indifference and preference thresholds specify the discriminating power of a criterion function and lead to the following definitions.

DEFINITION 9.3.2: A ***pseudo-criterion*** *is a criterion function g that is associated with two threshold functions $q_g[g(a)]$ and $p_g[g(a)]$ satisfying (see (r 9.3.10)):*

$$\forall\, a, b \in A, \quad \frac{q_g[g(a)] - q_g[g(b)]}{g(a) - g(b)} \geq -1, \quad \frac{p_g[g(a)] - p_g[g(b)]}{g(a) - g(b)} \geq -1,$$

and such that, $\forall\, a, a' \in A$:

$$g(a') \geq g(a) \Rightarrow \begin{cases} a'\, I_g\, a \text{ if } g(a') - g(a) \leq q_g[g(a)] \\ a'\, Q_g\, a \text{ if } q_g[g(a)] < g(a') - g(a) \leq p_g[g(a)], \\ a'\, P_g\, a \text{ if } p_g[g(a)] < g(a') - g(a). \end{cases}$$

Note that:

- the three "ifs" at the right of the implication can be replaced by "if and only if;"
- a true criterion is a pseudo-criterion such that $q_g[g(a)] = p_g[g(a)] = 0$, $\forall$ a $\in$ A.

DEFINITION 9.3.3: A pseudo-criterion g is called:
- *a* ***semi-criterion*** *when $q_g[g(a)] = p_g[g(a)]$, $\forall\, a \in A$,*
- *a* ***precriterion*** *when $q_g[g(a)] = 0$, $\forall\, a \in A$.*

Note that:

- there is no zone of weak preference with a semi-criterion: the two boundaries g(a) + $q_g[g(a)]$ and g(a) + $p_g[g(a)]$ become the same in Figure 9.3.4;
- there is indifference with a precriterion only when g(a′) = g(a): the two boundaries g(a) + $q_g[g(a)]$ and g(a) - $q'_g[g(a)]$ become the same as the diagonal in Figure 9.3.4, which would also be the case with a true criterion.

A pseudo-criterion g establishes a system of preference relations (I_g, P_g, Q_g), defined by Definition 9.3.2, on the set A. In practice, it is preferable to consider a pseudo-criterion as a functional representation of a system of preference relations along the significance axis g. In other words, in many cases the SPR (I_g, P_g, Q_g) represents the fundamental part of the preferences modeled by g, and the pseudo-criterion should be considered a convenient tool that can operate on the SPR rather than a precise instrument that assigns authoritative values to actions in A.

First, note that if g is a true criterion, the SPR (I_g, P_g) associated with it (Def. 9.3.1) is a complete preorder (section 7.2.2.2,a).

Considering result 7.2.2 from Section 7.2.2.2, one can argue that when the preference structure associated with the significance axis of criterion g is a complete preorder (I, P):

- either I and P coincide, respectively, with I_g and P_g associated with true criterion g, and g is thus a true criterion;
- or they do not, but one could find a nondecreasing monotonic transformation that changes g to g′ such that the relations $I_{g'}$ and $P_{g'}$ associated with the true criterion g coincide with I and P; a simple example would be when I and P are defined by

comparing the values of g after rounding them to the nearest integer – e.g., on the original criterion g, 3.8 and 3.9 might be considered indifferent, which would violate the complete preorder, but after rounding, their g' values are both 4, and the indifference would be compatible with the preorder on g'.

In the case where g is a semi-criterion, the unique threshold function can be denoted $q_g[g(a)]$. Based on Definition 9.3.2, the associated SPR is again of the form (I_g, P_g), which has the structure of a semi-order when $q_g[g(a)]$ is a constant q (see Section 7.2.2.2,b).

When g is some pseudo-criterion, Definition 9.3.2 implies that the associated SPR is of the form (I_g, P_g, Q_g). In the special case where the indifference threshold is a constant q, the SPR has the structure of a pseudo-order (Section 7.2.2.3,b). This result holds in the general case, as well.[17]

We summarize in the following result.

RESULT 9.3.1: Let g be a pseudo-criterion. The SPR (I_g, P_g, Q_g) that it defines has the structure of a pseudo-order, which becomes a semi-order when g is a semi-criterion and a directed semi-order when g is a pre-criterion.

9.3.4 Determining indifference and preference thresholds

a) General remarks

One would like to determine the values of indifference and preference thresholds through rigorous reasoning based on the components γ_i and δ_i forming the support of g and on the process that determines the criterion value from these components. Although the analyst will often find such an approach too difficult – for example, consider the time savings criterion in Example 2 (see Section 9.2.2.1,b) or the cost of reaching a thousand targeted readers in Example 5 (see Sections 8.1.5 and 8.2.2.1) – this subsection considers the cases where it is not.

The pseudo-criterion model is designed to account for the fuzziness that results from the imprecision of the components of the support of g and from the arbitrary components of the functional definition of g. It is thus intended to delimit **indifference and strict preference situations that are subject to as little debate as possible**. For example, consider the case where g is the mathematical expectation of a utility (Section 9.2.2.1, a2). Difficulties associated with the choice of probability distribution or the means of constructing the utility function for the degrees might cause one to question the appropriateness of the true-criterion model. Of course, these difficulties would also prohibit the analyst from deducing objective values for the discrimination thresholds. This does not mean, however, that one should choose $q_g = 0$, $p_g = 0$, which is implicitly

[17] *translator's note: Additional discussion is provided on pages 264 and 265 of the original, French version.*

done in traditional utility theory. A little common sense, a practical understanding of the orders of magnitude, a bit of trial and error, or some quick calculations may often lead to functions q_g and p_g that transform the mathematical expectation g into a pseudo-criterion that models preferences with much more realism than the true criterion of the classical theory.

The analyst will often have good reasons to consider these thresholds constant or proportional to g(a), or perhaps to be of the form $\alpha + \beta\, g(a)$. To determine the values of α and β, he could often follow the same reasoning used in Section 9.3.2 to define thresholds as upper and lower limits of a difference $g(a') - g(a)$. He could complement this reasoning with some quick calculations, simple sensitivity analyses, reflections on "intangibles," or changes in structural hypotheses. Readers can convince themselves that this type of approach can be very useful for criteria such as the cost in the highway location example and the total cost or the fracture stress in the Product Composition example (Section 9.1.1, Example 11). Similarly, the analyst could consider building a pseudo-criterion based on a lexicographic principle.[18]

b) Case where g's support is a point indicator with thresholds

When $\gamma_i(a)$ is the only component in the support of g and the positive and negative intrinsic dispersion thresholds (Section 8.2.2) have been defined, one can show that the discrimination thresholds can be determined from simple calculations that do not require additional data.[19]

c) Case where g is a point reduction criteria on dimension i[20]

9.4 GRADATIONS AND MEASURES

The concepts addressed in this section are essential for understanding how a criterion function g can be used to compare the preference differences or exchanges discussed in Section 7.2.4. It should be clear that difference comparisons based on the values of g for actions of A cannot be comprehensive if $\upsilon_g(A) \neq \upsilon(A)$ (Section 9.1.2,f). The comparisons would have to be restricted to g's significance axis in these cases. This is why we denote the indifference, preference, incomparability, and outranking relations (on A × A) studied in this section by I_g^*, P_g^*, R_g^*, S_g^*. Though not often discussed

---

[18] *translator's note: The original, French version uses the Engine Assignment Example (Example 9) to illustrate this approach on Pages 266-267.*

[19] *translator's note: The original, French version demonstrates this on Pages 268-270 (see also, for illustration on real-world problems, Roy et al., 1986, and Roy and Bouyssou, 1986).*

[20] *The original, French version introduces this case and illustrates it with the continuation of Examples 3, 6, and 7 on Pages 270-274.*

explicitly, these relations are important in constructing comprehensive preferences on A when additional criteria are relevant to these preferences (Section 11.1).

The subject has rarely been addressed as it is here,[21] and the questions raised in this section are just as important as their answers. We raise fundamental questions about the type of information contained in the criteria and the impact that this information can have on modeling comprehensive preferences.

Since the subject is much broader than it may seem, we have limited ourselves in several places to sketches based on examples or to difficulties that may arise. That is, the pioneering and incomplete nature of our research in this area, coupled with a concern for brevity, render a few of the subsections incomplete, and we apologize in advance.

9.4.1 Comparing preference differences along a criterion's significance axis

When a criterion g is intended to reflect comprehensive preferences on A by itself, comparing preference differences along g's significance axis amounts to comparing these same differences from a comprehensive point of view (Section 7.2.4.1); i.e., $I^*_g = I^*$, $P^*_g = P^*$, ... When there are other criteria relevant to the formation of comprehensive preferences, however, it is interesting to consider the relationship between comparing preference differences from a comprehensive perspective and comparing preference differences restricted to g's significance axis. By definition, all other aspects of the consequences not modeled in the support of g are ignored when forming these restricted comparisons. That is, the actions are considered equivalent on all criteria other than g.

We shall use an example to illustrate the issues involved, then specify in a synthetical way the nature of the relationship that guarantees the compatibility of a BSPR on $A \times A$ and each of the nine (basic and consolidated) relations that compare preference differences restricted to g's axis. After that, we present an interesting aspect that can arise when these relations are used to construct or discuss comprehensive preferences when there are several relevant criteria.[22]

Example 10: Application Packages (from section 7.2.4.1)

Assume that one criterion per subject is being considered to account for the scholastic results in the application packages. For each discipline, the criterion specifies the rules for aggregating the different results obtained so as to give a final grade between 0 and 20. For example, these rules may lead to weighting the results of some exams differently

---

[21] *For a classical approach, see Suppes and Winer (1955); Adams (1965); Fishburn (1970); Vansnick (1984).*

[22] *translator's note: Additional details are presented on Pages 276, 279, and 280 in the original, French version.*

than those of others, giving special importance to the best and worst grades, and so on. Consider g, here, to be the mathematics criterion obtained in this way.

Let us think again of the three candidates a, b, and c, and consider that the actor Z ranks a preferred to b preferred to c. She also considers b to be closer to c than to a; i.e., $(b \to a)\ P^*\ (c \to b)$. If one wants to say that such a comparison of preference differences takes place only along the significance axis of g, one must assume that this criterion is the only one responsible for the differences between a and b and between b and c. This would imply that a, b, and c appear as equivalent on every criterion other than g. Assume that this is indeed the case and that:

$$g(a) = 18,\ g(b) = 13,\ g(c) = 10.$$

Asserting that $(b \to a)\ P^*\ (c \to b)$ means that according to Z, the difference $18 \ominus 13$ is more important than the difference $13 \ominus 10$ for criterion g when considering candidates with the same results as a, b, and c on the other criteria. Depending on the influence that the values of these other common results could have on the preferences of Z, one may or may not be able to conclude $(b \to a)\ P^*_g\ (c \to b)$. Let a′, b′, and c′ be three new real or dummy candidates such that:

$$g(a') = 18,\ g(b') = 13,\ g(c') = 10,$$

where g is the only criterion that differs among these three candidates. If Z believes that $(b' \to a')\ P^*\ (c' \to b')$, no matter what the values of the common grades on the criteria other than g, then the statement $(b \to a)\ P^*\ (c \to b)$ is not based on the values of these other criteria and can be ascribed entirely to the significance axis of g. In this case, and only in this case, we write $(b \to a)\ P^*_g\ (c \to b)$.

Assume now that Z's BSPR on A × A is different from the one assumed above, so that considering candidates a′, b′, and c′ leads to:

$$(b' \to a')\ P^*\ (c' \to b') \text{ or } (b' \to a')\ Q\ (c' \to b').$$

The statement $(b \to a)\ P^*\ (c \to b)$ no longer reflects a comparison of intrinsic preference differences on the axis of criterion g in this case. This is why we would reject $(b \to a)\ P^*_g\ (c \to b)$ here. We would, however, accept $(b \to a) \succ^*_g (c \to b)$, since such an assertion is not based on the values of a, b, and c on criteria other than g. Note that, in this example, no matter what the common values of a′, b′, and c′ on criteria other than g, the weak and strict preferences between $(b' \to a')$ and $(c' \to b')$ occur in the same direction. If $(b' \to a')\ I^*_g\ (c' \to b')$ were also possible, this would not create any contradiction either; it would lead to our adopting $(b \to a)\ S^*_g\ (c \to b)$.

Actually, as we shall now show, it is quite possible to have "contradictory" exchanges. For this demonstration, consider the candidates with characteristics given in Table 9.4.1. Without demonstrating any incompatibility, Z might believe that:

- when faced with poor grades, increasing the mathematics grade from 10 to 13 would correspond to an exchange which she would value less than an exchange involving an increase in the same mathematics grade from 15 to 18;
- when faced with average grades, on the other hand, she would consider the two exchanges of equal value;
- when faced with excellent grades, she would consider the exchange (10 → 13) of significantly greater value than the exchange (15 → 18).

Notationally, we would write: (b → a) P^* (d → c); (b′ → a′) I^* (d′ → c′); (d″ → c″) P^* (b″ → a″). If, as before, we wish to restrict the comparisons to g's significance axis, we would only be able to conclude: (b → a) R_g^* (d → c); (b′ → a′) R_g^* (d′ → c′); (b″ → a″) R_g^* (d″ → c″), since comparing only the values 10 ⊖ 13 and 15 ⊖ 18 leads to contradictory strict preferences – i.e., a reversal of preference direction – depending on the values of the other criteria. Finally, consider the case in which no matter what these other values are, Z would never arrive at (d″ → c″) P^* (b″ → a″), but only at (d″ → c″) Q^* (b″ → a″). We would, thus, observe contradictory exchanges, but strict preferences would only occur in a single direction. We indicate this fact by replacing the relation R_g^* by K_g^* in this case.

Table 9.4.1: Characteristics of the candidates in Example 10

Candidates	g	Other criteria
a	18	No difference among a, b, c, d: poor grades on each criterion.
b	15	
c	13	
d	10	
a′	18	No difference among a′, b′, c′, d′: average grades on each criterion.
b′	15	
c′	13	
d′	10	
a″	18	No difference among a″, b″, c″, d″: excellent grades on each criterion.
b″	15	
c″	13	
d″	10	

(end of Example 10)[23]

This example shows that when comparing exchanges of actions differing only on their evaluations in criterion g, the comparisons must be made two by two in order to give meaning to comparisons of preference differences along the significance axis of g.

[23] *translator's note: Example 10 was continued and ended in Section 11.3.1 in the original, French version.*

Under these conditions, one may wonder about the stability of such comparisons when the evaluations of the four actions with respect to criteria other than g vary in all possible ways. We assume that the answers are determined by referring to a BSPR on A × A, either because it is known or because one is trying to construct it. Whatever the variations in the nature and the directions of the relations that affect these answers, it is always possible to summarize them with the help of one and only one of the nine following relations by conforming to the consistency principle sketched out below.[24] The rules governing the choice of which relation would be adopted in the two cases where we have grouped three relations together are those outlined in the framework of the preceding example.

When there is total stability in the comparisons	$(b \to a)\ I_g^*\ (d \to c)$ $(b \to a)\ P_g^*\ (d \to c)$ $(b \to a)\ Q_g^*\ (d \to c)$
When there are no contradictory exchanges in the comparisons	$(b \to a) \succ_g^* (d \to c)$ $(b \to a)\ J_g^*\ (d \to c)$ $(b \to a)\ S_g^*\ (d \to c)$
When there are contradictory exchanges but not related to strict preference	$(b \to a) \sim_g^* (d \to c)$
When there are contradictory exchanges and strict preference for (b, a) but not for (d, c)	$(b \to a)\ K_g^*\ (d \to c)$
When there are contradictory exchanges with preferences either for (b, a) or for (d, c)	$(b \to a)\ R_g^*\ (d \to c)$

Before proceeding, let us briefly indicate how the comparison of preference differences along a criterion g's significance axis can play a role in constructing comprehensive preferences. Consider four actions a, b, c, and d that are equivalent on all criteria other than g and such that:

$$(b \to a)\ S_g^*\ (d \to c).$$

Let h be a criterion different from g, and a′, c′ be two actions equivalent to a, b, c, and d on all criteria except g and h. Let these actions be such that:

$$g(a') = g(a) = x > g(b) = y;$$
$$g(c') = g(c) = z > g(d) = t;$$
$$h(a') = h(c') = \varepsilon < h(a) = h(b) = h(c) = h(d) = e.$$

Finally, assume b P a′.

---

[24] *translator's note: Remarks concerning these relations are presented on Page 279 in the original, French version.*

Note that moving from b to a′ constitutes an exchange that leaves all the evaluations on criteria other than g and h invariant, that deteriorates the evaluations along h from e to ε, and that improves the evaluations along g from y to x. This improvement is not enough to compensate for the deterioration along h, since b P a′. Note also that moving from d to c′ leads to the same observations, except that the improvement along the criterion g is from t to z instead of from y to x. Now, we have set (b → a) S_g^* (d → c). That is, the improvement in the evaluation of g from t to z does not lead to an increase in satisfaction that is greater than that brought about by the increase from y to x; recall that the evaluations along the other criteria remain constant. Under these conditions, the improvement from t to z cannot compensate for the deterioration from e to ε, since this deterioration cannot be compensated by the improvement from y to x. One must thus conclude d P c′ no matter what the common values of these two actions are for criteria other than g and h.

Let H_g^* denote any of the nine relations considered. One can always consider relations H_g^* as being defined on g(A) × g(A) and write:

$$(y \rightarrow x)\ H_g^*\ (t \rightarrow z);\ \text{or}\ (x \ominus y)\ H_g^*\ (z \ominus t),$$

to indicate that for the four actions a, b, c, and d equivalent in all criteria except g and such that:

$$g(a) = x;\ g(b) = y;\ g(c) = z;\ g(d) = t;$$

the statement (b → a) H_g^* (d → c) is true.

This notation has the advantage of indicating the values x, y, z, t of the criterion, without having to refer to the specific actions a, b, c, and d.[25]

9.4.2 Gradation and gradable criteria

Let x, y, z, and t be four performance levels on a criterion g. This subsection addresses the following question: How can comparing the numbers x − y and z − t be used to compare preference differences (x ⊖ y) and (z ⊖ t)?[26] The examples that follow illustrate the scope of this question and motivate subsequent definitions.

First, assume that the actions can be compared according to a single criterion g representing a monetary gain. Consider the case x − y = z − t = s > 0. Can one conclude (x ⊖ y) I_g^* (z ⊖ t)? Such an indifference would signify that the additional gain s has the same attraction when it is obtained from an exchange for which the original action is

[25] *translator's note: The disadvantage is shown in a brief discussion presented on Page 281 of the original, French version.*

[26] *Consistent with the notation used at the end of Section 9.4.1, the relation ⊖ is used with performance levels.*

evaluated at y as when it is obtained from an exchange for which the original action is evaluated at t. This may be so in many cases, but it is also possible that some actor considers the attraction of the extra s to decrease (or increase) as a function of the quantity to which it is added.

Now think of a, b, c, d, ... as being musical recordings on magnetic tapes, where the tapes have all been wound on identical reels. Let x, y, z, t, ..., respectively, be the number of revolutions recorded on each reel, and assume that when all other elements are the same, longer recordings are preferred to shorter ones. Such a criterion (the x, y, z, t, ...) is not well suited to comparing preference differences, since the playing time in a revolution becomes longer as the distance from the center increases; i.e., the increased playing time in an extra $s = x - y$ revolutions increases as y increases.

Notice that the number $x - y$ and $z - t$ can always be compared in terms of numbers. From this fact, the question raised at the beginning of this subsection can only have a simple answer if the three relations R^*_g, $\sim^*_g$, K^*_g are empty on $g(A) \times g(A)$. **Therefore, from here on we shall only be interested in criteria g such that, whatever the values of x, y, z, t, at least one of the following can be considered true**:

$$(x \ominus y)\ S^*_g\ (z \ominus t);\ (z \ominus t)\ S^*_g\ (x \ominus y).$$

It should be clear that there will be an answer satisfying the question raised if:

$$x - y \geq z - t \Rightarrow (x - y)\ S^*_g\ (z - t) \qquad \text{(r 9.4.1)}$$

The implication going in the other direction could be violated, however.

Example 12: Plant Organization (from Section 7.2.4.1)

Let g be a criterion for which the significance axis is "adaptability to disruptions." The axis attempts to reflect the plant's ability to remain productive under a given organization scheme during variations in anticipated supply and when faced with short-term fluctuations in production requirements. To define g(a) for the different types of organizations considered, assume that the analyst questions the different stakeholders about the five reference actions defined in Section 6.2.2.

Suppose that all agree to accept that:

- a_1 is the most rigid type of organization;
- a_2 is definitely more flexible than a_1, but still is quite rigid;
- a_3 offers an ability to adapt to disruptions that is definitely superior to that of a_2;
- a_4 and a_5 correspond to types of organizations that are difficult to differentiate along g's significance axis and for which the abilities to adapt to disruptions are among the most advanced that one could imagine.

The analyst thinks that these reference actions will allow him to characterize benchmark points on the significance axis of g that he can use to situate other types of organizations on this axis. With this in mind, he questions the various stakeholders to compare the preference differences separating the benchmark levels when ignoring all other considerations. Assume that there is a consensus to accept that:

- the increased adaptability when passing from a_1 to a_2 is significantly smaller than that obtained when passing from a_2 to a_3, which is itself smaller than that obtained when passing from a_3 to a_4;
- the increased adaptability when passing from a_1 to a_3, however, is larger than that passing from a_3 to a_4.

The analyst believes that, under these conditions, he can say:

$$(g(a_1) \rightarrow g(a_4)) \succ_g^* (g(a_2) \rightarrow g(a_4)) \succ_g^* (g(a_1) \rightarrow g(a_3)) \succ_g^* (g(a_3) \rightarrow (a_4)) \succ_g^* (g(a_2) \rightarrow g(a_3)) \succ_g^* (g(a_1) \rightarrow (a_2)). \quad \text{(r 9.4.2)}$$

He believes that the first two comparisons in this chain follow logically from the other comparisons that model the information he has obtained.

To facilitate the discussion, he finds it convenient to assign a scale to g ranging from $g(a_1) = 0$ to $g(a_4) = g(a_5) = 20$. If he sets $g(a_2) = 6$ and $g(a_3) = 13$, he notices that:

$$g(a_4) - g(a_3) = 20 - 13 = 13 - 6 = g(a_3) - g(a_2).$$

These preference differences are not indifferent, however, since (r 9.4.2) says that $(g(a_3) \rightarrow g(a_4))$ is preferred to $(g(a_2) \rightarrow g(a_3))$. The relation S_g^* defined by (r 9.4.1), therefore, does not coincide with the model of (r 9.4.2). To remedy the problem, he could, for example, keep $g(a_2) = 6$ and set $12 < g(a_3) < 13$ (see Table 9.4.2, solution 2). There are other values of $g(a_2)$, of course, that would make (r 9.4.1) and (r 9.4.2) compatible (see Table 9.4.2, solutions 4 and 6). The possible choices are constrained, however, by the following easily derived conditions:

$$g(a_2) < 20/3; \; 10 < g(a_3) < 10 + 1/2 \times g(a_2).$$

Once the values of $g(a_2)$ and $g(a_3)$ are fixed, the analyst can refine this "calibration" of g's significance axis by proceeding in a similar fashion with a few other intermediate reference actions. He could then evaluate any action a by determining an interval around g(a) by using the g values of two consecutive benchmarks and then positioning g(a) in the interval by comparing action a to the actions corresponding to the bounds on this interval. He would attempt to verify that (r 9.4.1) would not be violated by any actions considered.

(end of Example 12)[27]

[27] *translator's note: Example 12 was continued and ended in Section 12.6 in the original, French version.*

Table 9.4.2: Examples of g-values for the reference actions of Example 12

Solution	$g(a_1)$	$g(a_2)$	$g(a_3)$	$g(a_4)$
No. 1	0	6	13	20
No. 2	0	6	12.5	20
No. 3	0	5	10	20
No. 4	0	5	11	20
No. 5	0	3	9	20
No. 6	0	3	11	20

Only solutions 2, 4, and 6 make (r 9.4.1) and (r 9.4.2) compatible.

DEFINITION 9.4.1: A criterion function g is a ***gradation*** *relative to a set A if and only if the function* $g^*(a' \ominus a) = g(a') - g(a)$ *can be considered a criterion on* $g(A) \times g(A)$ *that allows preference difference comparisons along the significance axis of g. This is equivalent to:*
$g(a') - g(a) \geq g(b') - g(b) \Leftrightarrow [g(a) \ominus g(a')]\ S_g^*\ [g(b) \ominus g(b')]$.

The process for constructing g(a) indicated above in Example 12 is designed to make a gradation out of this criterion. In the example of musical recordings on magnetic tapes, knowing the thickness of the tape and the diameter of the spool of the reel, we could calculate the length $\chi(x)$ of a recording with x revolutions; if we consider that an extra minute of recording is roughly of equal value whether it is added to a short or long recording, then $\chi(g)$ is a gradation, even though g is not. Note here that χ is an encoding defined on g(A). The reader might wonder in what cases the monetary gain criterion discussed at the beginning of Section 9.4.2 is a gradation and, when it is not, if an encoding χ on g(A) can be found so that $\chi(g)$ is a gradation.

DEFINITION 9.4.2: The significance axis of a criterion g (or, roughly speaking, the criterion g) is considered ***gradable*** *relative to A when an encoding* χ *defined on g(A) exists such that* $\chi(g)$ *is a gradation.*[28]

9.4.3 Measures: Preference difference commensurability along a criterion's significance axis

We mentioned in Section 9.4.1 that comparing preference differences along g's significance axis is closely related to comparing specific exchanges of $A \times A$. These exchanges involve four actions that are considered equivalent on all criteria other than g. From now on we denote the set of quadruplets of actions with these properties $[A^4]_g$.

[28] *translator's note: The original French version discusses this definition on Pages 285-287.*

To give meaning to such comparisons and to facilitate qualitative preference responses, it is helpful to use a dimension that is external to the problem. This dimension might be, for example, a level of satisfaction, a willingness to pay a monetary sum, or a willingness to consume time in uninteresting work (see Section 7.2.4.1,a). To compare two exchanges, the values of the criterion on this exterior dimension are varied. The external dimension is defined so that these variations are comparable. In certain cases, one might be tempted to go further than simple qualitative comparisons. The exterior dimension can indeed allow one to assign some significance to the measure of the preference difference $a \ominus b$ (or to the exchange $(b \rightarrow a)$) when using some preference difference $c \ominus d$ (or the exchange $(d \rightarrow c)$) as the unit measure. This can be thought of as the ratio of preference differences $a \ominus b$ to $c \ominus d$, which can, in general, be denoted $r(\frac{a \ominus b}{c \ominus d})$, a real valued function defined on A^4.

Here, we are interested in the restrictions on this function $r(\frac{a \ominus b}{c \ominus d})$, in the set $[A^4]_g$, and in the relationships between it and the criterion function g when g is a gradation. The reader should reconsider the examples of Sections 9.4.1 and 9.4.2 in light of these issues.

Consider again the criterion g, duration of a taped recording, presented at the beginning of Section 9.4.2. Assume that this is a gradation and that the following proposition seems realistic: "Everything else being equal, an increase of m minutes in the length of recording leads to an increase in satisfaction m times that which would result from an increase of one minute." In this case, we could compare preference differences in a stronger than qualitative way. We could restrict $r(\frac{a \ominus b}{c \ominus d})$ in $[A^4]$ by:

$$r(\frac{a \ominus b}{c \ominus d}) = \frac{g(a) - g(b)}{g(c) - g(d)}.$$

More generally, we say that **a gradation g conforms to the requirement of preference difference commensurability along its significance axis** if, for any quadruplet (a, b, c, d) $\in [A^4]_g$, the above equality seems a realistic way to restrict r in this set.

Consider two gradations g and $\chi(g)$ that have the same significance axis and that both conform to the preference difference commensurability requirement. It follows that:

$$r(\frac{a \ominus b}{c \ominus d}) = \frac{g(a) - g(b)}{g(c) - g(d)} = \frac{\chi(g(a)) - \chi(g(b))}{\chi(g(c)) - \chi(g(d))},$$

from which one can deduce:

$$\frac{\chi(g(a)) - \chi(g(b))}{g(a) - g(b)} = \alpha,$$ where α is a positive quantity that does not depend on a or b, and:

$$\chi(g(a)) - \alpha\, g(a) = \beta,$$ where β is a quantity that does not depend on a.

This establishes that the only encodings that transform a gradation conforming to the requirement of preference difference commensurability into another gradation also conforming to this requirement are affine transformations:

$$\chi(g(a)) = \alpha\, g(a) + \beta.$$

The preference difference commensurability requirement is not the only one we may wish to impose on a gradation, of course. It is not even the only requirement for which two gradations conforming to the property are related by an affine transformation.[29] It seems difficult to establish this result for more general conditions, however. This leads us to propose the following definition:

DEFINITION 9.4.3: We shall say that a requirement renders ***the significance axis*** *of a criterion g* ***measurable*** *if it is gradable and if the gradations conforming to this requirement are determined up to an affine transformation. Every gradation conforming to a requirement that renders the significance axis of a criterion g measurable will be called a* ***measure***.[30]

RESULT 9.4.1: Whatever the nature of the requirement that renders the significance axis of a criterion g measurable, the function:

$$r\left(\frac{a \ominus b}{c \ominus d}\right) = \frac{g(a) - g(b)}{g(c) - g(d)}$$

defined, $\forall\ (a, b, c, d) \in [A^4]_{g}$, *such that* $g(c) \neq g(d)$, *is independent of the measure g chosen and satisfies:*

$$r\left(\frac{a \ominus b}{c \ominus d}\right) = 1 \Rightarrow (a \ominus b)\ I_g^*\ (c \ominus d);$$

$$(a \ominus b)\ P_g^*\ (e \ominus e) \text{ and } (c \ominus d)\ P_g^*\ (f \ominus f) \Rightarrow r\left(\frac{a \ominus b}{c \ominus d}\right) > 0,\ r\left(\frac{a \ominus a}{c \ominus d}\right) = 0$$

$$r\left(\frac{a \ominus b}{e \ominus f}\right) + r\left(\frac{b \ominus c}{e \ominus f}\right) = r\left(\frac{a \ominus c}{e \ominus f}\right)$$

$$r\left(\frac{a \ominus b}{e \ominus f}\right) \times r\left(\frac{e \ominus f}{c \ominus d}\right) = r\left(\frac{a \ominus b}{c \ominus d}\right).$$

---

[29] *translator's note: An illustration presented on Pages 288-289 of the original French version shows that under certain conditions, gradations conforming to a property stating that thresholds can characterize the discriminating power of a criterion are related by affine transformation.*

[30] *translator's note: An additional requirement that makes the significance axis of a criterion measurable is presented on Page 209 of the original, French version of this book.*

When the requirement considered is different from that of commensurability, the function r defined above can be used as a basis of preference difference commensurability along g's significance axis.[31]

9.4.4 Von Neumann-Morgenstern expected utility criteria and preference difference commensurability based on lottery comparisons

9.4.4.1 Axiomatic foundations

Von Neumann-Morgenstern utility theory (see Von Neumann and Morgenstern, 1967; Raiffa, 1970 or de Neufville, 1991) considers the comparison of (ideal, see Section 9.1.2, remark b) actions from a set A that have been evaluated on a single dimension through a nonpoint state indicator $\gamma(a)$. The likelihoods of the states covered by the indicator are assumed to be captured exactly and nonarbitrarily by probabilities assigned to the states:

$\forall$ a $\in$ A, $\delta^a(e)$ is the known probability that degree e $\in \gamma(a)$ will result if a is executed.

What assumptions on Z's preferences need to be made so that a true criterion $g^u(a)$ of the form:

$$g^u(a) = \sum_{e \in \gamma(a)} u(e) \cdot \delta^a(e),$$

can be used to account for these preferences? The criterion $g^u(a)$ is the mathematical expectation of the utilities u(e) associated with the degrees of the scale. We already saw this type of criterion in Section 9.2.2.1,a2. Once again, we use e_* and e^* to denote the extreme degrees of the scale E being considered.

Utility theory is based on the following four axioms.

AXIOM 9.1 (no ambiguity in comparisons): $\forall$ *a, b* $\in$ *A characterized by probability functions* δ^a *and* δ^b*, one and only one of the following possibilities can hold:*
a I b (and b I a); a P b; b P a.

That is, using a criterion that summarizes the information relative to the dimension considered leads to refusing any incomparability. Axiom 9.1 goes even further, since it excludes any ambiguity or hesitation between situations of indifference and strict preference.

AXIOM 9.2 (transitivity): The binary relations I and P are transitive.

[31] *translator's note: The implications of Result 9.4.1 are discussed in more detail on Pages 290-292 of the original, French version. In that version, the result corresponds to RÉSULTAT 9.4.2, since RÉSULTAT 9.4.1 was not translated in the English version.*

In the context of the problem we are considering, Axioms 9.1 and 9.2 imply that Z's preferences are compatible with the axiom of complete transitive comparability 7.1.2.

The two other axioms introduce actions (which can always be assumed to belong to A) characterized by probability distributions derived from those of two other actions by a random drawing according to given probabilities π and 1 – π. Specifically, given two actions a′ and a such that a′ P a and $\pi \in \mathbb{R}$ with $0 \leq \pi \leq 1$, we denote:

l(π) the ideal action characterized by the probability distribution

$$\delta^{l}(\pi) = (1 - \pi)\, \delta^{a} + \pi\, \delta^{a'}.$$

This can be interpreted as a **lottery** with actions a′ and a that would obtain with probabilities π and 1 – π, respectively. The set of such lotteries will be denoted L(a, a′).

AXIOM 9.3 (continuity): ∀ a, b, a′ ∈ A such that a′ P b and b P a, there is a single value of π (0 ≤ π ≤ 1) for which the lottery l(π) ∈ L(a, a′) is indifferent to b.

The procedure outlined at the end of Section 9.2.2.1a2 to construct the utility function u(e) (which we denoted v(e) there) is based on this third axiom. In fact, let the two actions characterized by the degenerate probability distributions concentrated at e_* and e^*, respectively, correspond to a and a′, and let us call **basic lotteries** those lotteries in the set L(a, a′) that result. The procedure consists of comparing basic lotteries l(π) to actions b(e) characterized by degenerate probability distributions concentrated at state e. The objective is to determine a sufficient number of couples (π, e) that represent the indifference l(π) I b(e). For any degree e different from e_* and e^*, we have a′ P b(e) and b(e) P a and, according to the axiom, there exists a unique value of the probability π that is compatible with l(π) I b(e). In practice, this value of π may be poorly determined over a fairly large interval.

Note that the use of hypothetical basic lotteries gives meaning to the function u(e) (and, therefore, to a criterion $g^{u}(a)$), even when all of the actions in A have certain deterministic consequences – i.e., when γ(a) is a point estimate, ∀ a ∈ A.

AXIOM 9.4 (substitution or independence): Consider two ideal actions b and b′, respectively, characterized by:

$\delta^{b} = \pi\, \delta^{c} + (1 - \pi)\, \delta^{d}$;

$\delta^{b'} = \pi\, \delta^{c'} + (1 - \pi)\, \delta^{d}$.

∀ c, c′, and d ∈ A, and 0 < π ≤ 1, we have:

c′ I c ⇔ b′ I b, and c′ P c ⇔ b′ P b.

This axiom expresses the fact that if some action c′ is substituted for the action c in a lottery like b, the resulting preference when comparing b and b′ is the same as that when comparing c and c′. The example of the mayor's preferences (see Section 7.2.1.2) can be used to illustrate the implications of this axiom.

Let us ignore the costs and compare the projects only according to the number of jobs that they create. Consider an additional project d that would create 10 jobs with certainty. It follows from Table 7.2.1 that:

$$\delta^{a^1} = p\ \delta^{a^3} + (1 - p)\ \delta^{d}.$$

Let a^1 and a^3 take the place of b and c, respectively, in Axiom 9.4. Substituting a^4 for a^3 ($a^4 = c'$), we get:

$$\delta^{b'} = p\ \delta^{a^4} + (1 - p)\ \delta^{d} = \delta^{a^2}.$$

That is, a^2 takes the place of b′. Axiom 9.4 implies that (ignoring cost considerations):

$$a^1\ I\ a^2 \Leftrightarrow a^3\ I\ a^4, \text{ and } a^1\ P\ a^2 \Leftrightarrow a^3\ P\ a^4.$$

As we have seen, however, it is perfectly reasonable to accept $a^2\ P\ a^1$ and $a^3\ P\ a^4$, which would violate axiom 9.4.

The fundamental result of von Neumann-Morgenstern utility theory is the following:

RESULT 9.4.2: If the preferences of an actor Z conform to Axioms 9.1, 9.2, 9.3, 9.4, then they can be modeled by a true criterion g^u, defined up to a positive affine transformation, such that the value of $g^u(a)$ is obtained by taking the mathematical expectation of the utilities associated with action a.

This result can be fairly easily derived (see Raiffa, 1970) by using the procedure for constructing the function u(e) described above. Consider this procedure once again, but applied to the dimension "number of jobs created by a project" in the context of the example of the mayor's preferences. Notice that the (e, π) pairs compatible with the indifference statement b(e) I l(π) has meaning, even for those not willing to accept Axioms 9.1 to 9.4.

Set:

$$e_* = 10,\ u(10) = 0;\ e^* = 110,\ u(110) = 1.$$

Assume that an individual accepts indifference for the following (e, π) pairs:

$$(20, 0.2);\ (50, 0.55);\ (70, 0.75);\ (100, 0.95).$$

This would lead to the curve in Figure 9.4.1. Note the concavity of this curve. This is interpreted as aversion toward risk: A project creating n jobs with certainty is preferred to another creating n + 1 jobs with 50 chances out of 100 and n − 1 jobs with 50 chances out of 100, whatever the value of n.

After not having considered this curve for some time, the individual furnishing the (e, π) pairs above could look for pairs such that b(e) I l(π). But this time, instead of furnishing values in response to given e values and reasoning only on the basis of basic lotteries, she would only use 50-50 lotteries (i.e., π = 50/100) that would call into play degrees other than e_* and e^* in the lotteries and furnish their certainty equivalents e. This new procedure might, for example, lead to the following results:

First 50-50 lottery: (110, 10); Result: e = 45, which implies u(45) = (1 + 0)/2 = 1/2.
Second 50-50 lottery: (110, 45); Result: e = 70, which implies u(70) = (1 + 1/2)/2 = 3/4.
Third 50-50 lottery: (45, 10); Result: e = 23, which implies u(23) = (1/2 + 0)/2 = 1/4.
Fourth 50-50 lottery: (110, 70); Result: e = ...

Figure 9.4.1: Illustrative utility function relative to the "jobs created" dimension in the example of the mayor's preferences

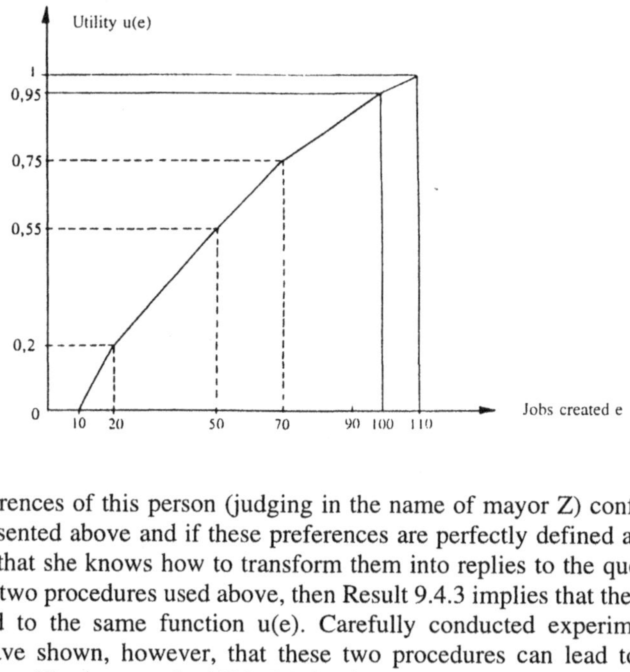

If the preferences of this person (judging in the name of mayor Z) conform to the four axioms presented above and if these preferences are perfectly defined and stable in her mind such that she knows how to transform them into replies to the questions posed in each of the two procedures used above, then Result 9.4.3 implies that the two procedures should lead to the same function u(e). Carefully conducted experiments in various contexts have shown, however, that these two procedures can lead to very different functions (see Hershey *et al.*, 1982; McCord, 1983; Bouyssou, 1984; McCord and de Neufville, 1984; Jaffray and Cohen, 1985).

From here on, we shall consider the case where the function u(e) is assumed to be perfectly defined (up to an affine transformation) by, for example, one or the other of the above procedures which are hypothetically assumed to be equivalent. One might wonder if this utility function can furnish any information with respect to the comparison of preference differences: Is an observed indifference between the certain creation of 45 jobs and the equally likely creation of 10 or 110 jobs logically equivalent to an indifference between the following two exchanges – passing from 10 to 45 jobs and passing from 45 to 110 jobs? Likewise, do the equal utility differences obtained when passing from 10 to 23 jobs, from 23 to 45 jobs, from 45 to 70 jobs, from 70 to

110 jobs reflect an indifference among the corresponding preference differences? We finish this chapter by addressing this issue.

9.4.4.2 Expected utility as a measure

The answer to this question is yes if and only if the expected utility criterion g^u is a gradation. Indeed, if it is not a gradation, g^u cannot be a measure (see Def. 9.4.3). On the other hand, if g^u is a gradation, then Result 9.4.3 implies that every criterion fulfilling the requirements of Axioms 9.1 to 9.4 is defined to an affine transformation. These requirements – i.e., these axioms – render the significance axis of g^u measurable.[32]

When Axiom 9.5 below is added to the four previous axioms and a few additional weak restrictions are added, g^u becomes a gradation and, therefore, a measure.

AXIOM 9.5 (comparability of preference differences using probabilities): $\forall\, \pi, \pi', \lambda, \lambda' \in [0, 1]$, *the preference differences between basic lotteries* $l(\pi)$, $l(\pi')$, $l(\lambda)$, $l(\lambda')$ *satisfy:*
$(l(\pi') \ominus l(\pi))\ I_g^*\ (l(\lambda') \ominus l(\lambda)) \Leftrightarrow \pi' - \pi = \lambda' - \lambda$,
$(l(\pi') \ominus l(\pi))\ P_g^*\ (l(\lambda') \ominus l(\lambda)) \Leftrightarrow \pi' - \pi > \lambda' - \lambda$.

We return to the example of the mayor's preferences to see the meaning of this axiom. Let a and b be two projects considered by the town whose consequences in terms of jobs are given in Table 9.4.3. Assume that the mayor can do something to change either a to a′ or b to b′, where these changes would only affect the consequences on jobs in the way shown in Table 9.4.3 (with $0 < r < 0.9$).

Table 9.4.3: Characteristics of (ideal) actions a, a′, b, b′

Projects	Probability of creating 10 jobs	Probability of creating 110 jobs
a	1.0	0.0
a′	0.85	0.15
b	1 – r	r
b′	0.9 – r	r + 0.1

To say that Z's preferences conform to Axiom 9.5 means that Z would rather change a to a′ than b to b′, since the preference differences $(a' \ominus a)$ and $(b' \ominus b)$ are 0.15 and 0.10, respectively, for any value of r.

Assume that g^u is a measure (or simply that g^{u*} is a true criterion), and consider three ideal actions a, a′, a″ such that the comparison corresponds exactly to that of the degrees:

---

[32] *translator's note: On Page 297, the original, French version shows that an additional axiom, which can take several forms, is necessary to consider g^u as a gradation. The presentation leads to Axiom 9.5.*

$$e = \gamma(a),\ e' = \gamma(a'),\ e'' = \gamma(a''),\ \text{with } e'' < e < e'.$$

Assume that these actions have been chosen so that the preference differences ($a' \ominus a$) and ($a \ominus a''$) are indifferent. Since g^u is a measure:

$$g^u(a') - g^u(a) = g^u(a) - g^u(a'') = u(e') - u(e) = u(e) - u(e'').$$

This leads to:

$$u(e) = \frac{u(e') + u(e'')}{2}.$$

This shows that the ideal action b corresponding to the lottery offering actions a′ and a″ each with probability 1/2 is indifferent to a. Moreover:

$$a\ I\ b \Rightarrow (a' \ominus a)\ I_g^*\ (a \ominus a'').$$

From this, and with the addition of weak additional hypotheses,[33] one should be able to deduce that the following axiom makes g^u a gradation and, therefore, a measure (as in Axiom 9.5).

AXIOM 9.6 (indifference between preference differences using equal probabilities): Whatever the four ideal actions a, a′, a″, b with evaluations:
$\gamma(a) = e,\ \gamma(a') = e',\ \gamma(a'') = e''$,
$\gamma(b) = \{e', e''\},\ \delta^b(e') = 1/2,\ \delta^b(e'') = 1/2$,
we have $a\ I\ b \Leftrightarrow (a' \ominus a)\ I_g^*\ (a \ominus a'')$.

Like the preceding axiom, this one uses probabilistic considerations as the basis of preference difference comparisons between ideal actions with point evaluations (i.e., actions determined exactly without any error, ambiguity, or uncertainty). The probability, in a sense, serves as a reference dimension by which to gauge the relative size of the preference differences considered.

---

[33] *The details can be found in Bouyssou (1984). One can also see Vansnick (1984).*

LEVELS III AND IV

HOW TO PROCEED FROM MULTIPLE CRITERIA TO COMPREHENSIVE PREFERENCES AND DEVELOP RECOMMENDATIONS

This last part presents a general description of how the analyst can use the concepts, results, and techniques developed at the other levels to answer the questions that have been posed and to aid the decision maker. It consists of an overview of the final elements of the proposed methodology: concepts, options, and procedures found at Levels III and IV (see Section 4.2). We consider it an overview, since it is presented for the most part in the form of reference examples and cannot, therefore, be considered systematic or complete. Many of the points not developed in detail will be covered in a second volume.

We remind the reader that Examples 1 to 12 are not treated as case studies (see Chapter 3). They have been introduced to illustrate concrete problems, approaches, and solutions. This means that we have taken the liberty of modifying the details of what truly happened in the original studies, which are quite old. Similarly, we use them to emphasize ways to proceed, suggestions, and notions that may not have arisen until after the original studies were completed. In the second volume mentioned above, true cases will be presented in a different fashion, one which shows in detail the operational nature of concepts, models, and procedures too difficult, intricate, or problem-specific to present here.

Chapter 10

COHERENT CRITERION FAMILY AND DECISION AIDING IN THE DESCRIPTION PROBLEMATIC

SUMMARY

The model Γ(a) presented in Section 8.2.5 does not generally allow the comparison of two actions. Therefore, we use the techniques presented in Chapter 9 to synthesize Γ(a) into a criterion family F. In Section 10.1 we show that for both theoretical and practical reasons there are no set rules for automatically deducing F from Γ(A). However, the analyst must respect some logical requirements, which then lead to the definitions of exhaustiveness, cohesiveness, and nonredundancy that characterize the concept of a coherent criterion family.

In Section 10.2, we introduce the performance tableau which indicates the performance level for each criterion member g_j of a coherent family F for each action a in a subset A′ of A. Indifference and preference thresholds associated with the criteria can also be included in the tableau. In a description problematic, the performance tableau usually represents the final product of the study. We highlight the types of fruitful discussions these tableaus can engender and caution against their common misuses.

In Section 10.3 we discuss several forms of dependence among criteria and place them in two main categories. We also discuss these types of dependence in the context of the two major approaches to preference modeling. Section 10.3.1 presents these descriptive and constructive approaches. The former is based on the existence of a rational decision maker with a coherent and stable SPR that is to be described as reliably as possible. The latter pays special attention to the conflicting and unstable nature of preference judgments and emphasizes the importance of significance axes for facilitating discussion of these preferences and constructing one or several SPR's. In Section 10.3.2 we show how the components of the criterion supports – the state indicators, dispersion indicators, and factors used to define them – can cause dependence among criteria and emphasize the often contingent nature of the set A. Although in a descriptive approach this form of dependence will lead to a desire to reduce the number of criteria, it is not considered a weakness in a constructive approach. We turn our attention in Section 10.3.3 to dependence stemming from value systems. Such dependence can be characterized by the fact that one cannot reason on the basis "all other things considered equal." One type of this dependence is related to the very notion of a criterion, which implies a certain ability to consider the criterion in isolation from others in the family F and gives meaning to the idea of preferences restricted to a significance axis of the criterion. Such a dependence, called utility dependence, is extremely troublesome in a descriptive approach. In a constructive approach, utility dependence is considered to be the sign of a missing criterion. We then introduce a second way of reasoning based on "all other things being equal" for a subfamily J of F. This "preference independence" of J in F allows the possibility of replacing the multiple criteria of J by a single criterion.

In Section 10.4, we contrast multicriteria and single criterion analysis. Multicriteria analysis is based on value systems that make explicit a family F of n (n > 1) unanimous, clear, and exhaustive criteria. Single criterion analysis avoids such explicitness by amalgamating, often prematurely, two types of information – information related to the consequences of actions and intercriteria information that is strongly influenced by the actors' value systems. To finish, we discuss the notions of dominance, substitution rate, concordance, discordance, and veto in the context of interpreting performance tableaus in δ-problematics.

In this chapter we demonstrate the useful role that the analyst can play in the decision process when employing the concepts and results of the preceding chapters. There are many decision situations where these are sufficient, although some situations need more elaborate concepts, models, and techniques. These latter situations will inevitably require more information than that involved in the description problematic P.δ – information dealing with, for example, the relative importance of criteria, dependence among criteria, or the aggregation logic from which a comprehensive system of preference relations will be developed. Since subjective judgments are required to obtain such information, the model becomes intertwined with the value system of the decision maker, and it can no longer be considered to address the differences that may arise among the various stakeholders in an impartial way (see Chapters 2 and 4). We discuss these issues throughout the following three chapters, but they are less relevant in this chapter because of the limited objective of the problematic considered.

This chapter deals with decision aiding in the framework of problematic P.δ (description). Recall from Section 6.1.4 that this problematic is an essential component of each of the others. It, therefore, becomes extremely important to define and formulate the problem properly in terms acceptable to the various actors – in others words, to identify clearly the possibilities, effects, and attributes upon which the performances of actions will be assessed. The proposed methodology allows the effort to be conducted rigorously and systematically, yet in a fashion that is straightforward and understandable by most of the actors. As we shall see, it leads to defining a (coherent) family of criteria that allows the most pertinent evaluations to be synthesized into what we shall call a performance tableau. This tableau can play an important role in the decision process even without any further modeling. Some explanation and care on the analyst's part are required, however, as shown in the examples and topics covered in the last two sections.

In Chapter 11 we go beyond the description problematic and see how the analyst formally and explicitly approaches the difficult problem of aggregating performance levels. There, we present the three major types of approaches that cover the principal models and methods. Combining these operational approaches with problematics α, β and γ (choice, sorting, ranking) is then treated in the final chapter, where we discuss several special problems of general interest. Bibliographic references allow the reader to go deeper into specific points not covered in the present volume.

10.1 COHERENT CRITERION FAMILY

At this point, we are interested in operationally synthesizing the information associated with the consequences of action a that is considered useful in comparing this action with another. In Chapter 8 we defined the term consequences (Def. 8.1.1) and described a methodology that allows the associated information to be structured and analyzed in a rigorous fashion. This methodology leads to model Γ(a) (Section 8.2.5). In Chapter 9, we presented different techniques that exploit certain categories of the information contained in Γ(a) by encoding it in a criterion. Recall that the objective of a criterion is to structure the result of the comparisons on a well-defined significance axis associated with the category of consequences considered.

In some cases the support of the criterion will use all the information, rather than a particular category of it. In this case the criterion itself will be the desired synthesis. Consider again the example of Section 7.2.1.2, where the mayor was comparing various projects designed to reduce unemployment. Assume that the only consequence categories to be considered are those dealing with the number of jobs created by each project and the cost to the municipality, and that the task is to guide the mayor's reasoning through these two criteria. The objective of each criterion would be to synthesize the information relative to one of these categories, each of which represents a major aspect of the problem.

In many cases, it is not easy to develop a criterion family that sufficiently synthesizes all the evaluation information of model $\Gamma(a)$. The analyst will often have several possibilities, and his choice at this stage will be important. We shall say more about this later.

The reader may wonder whether some automatic rules could be applied to derive the criterion family from model $\Gamma(a)$. There are both theoretical and practical reasons why this cannot be done.

As shown in Sections 9.2.3 and 9.2.2.2, one cannot systematically associate a criterion with each of the $\bar{n}$ dimensions of the consequence spectrum. The relationships between the significance axis of a criterion and the dimensions implied in its support are often complex. Even when these relationships are simple, the choice of an encoding function (Section 9.2.1) and, especially, of a point operator (Section 9.2.2.1) cannot be decided on technical reasons alone. In short, from a theoretical perspective, it is important to realize that the significance axes will generally not be obvious to the analyst. Indeed, the nature of the critical point representing the end of the present phase of analysis, the personalities and logic of the principal actors, and the sociological aspects that affect the course of the decision process can all strongly influence the choice of these significance axes. Still, there will remain a great deal of flexibility when specifying certain axes or, on the other hand, when implicitly considering them indirectly through the use of other axes (see Section 10.3 below).

Even though the analytical effort advocated in Chapter 8 is quite important, the details and most of the results will soon be forgotten: Once the criterion family has been defined, it alone will form the basis for discussion, and the more basic elements will not even be familiar to those outside of the working group. From a practical perspective, then, it is essential that the criterion family be conceived in such a way that it is understandable to all users and that it produces as wide a consensus as possible. These two requirements directly impact the analysis and should, therefore, be considered carefully when defining the n significance axes or the characteristics of the n criterion functions associated with them.

The analyst, then, needs to make the most of the principles and techniques in Chapter 9 to build a criterion family that can be **understood and accepted by all those who are part of the decision process**. In particular, he must:

- give the utmost attention to the explicit or implicit significance axes that form the basis by which the preferences of the different actors are formed, thought through, and transformed; by this, we mean that not only must the essential aspects of the effects and attributes considered in Γ(a) be represented by the different criteria, but the support of each of these criteria must be formed from their distributions in such a way that the explicit or implicit significance axes be made as clear a possible;[1]

- avoid using parameters outside of Γ(a) when defining the criteria, parameters whose estimation or interpretation lead to divergent points of view that could significantly affect (with respect to the indifference and preference thresholds) the levels taken by the criteria; here, we are thinking, for example, of the relative importance of certain elementary consequences that might be perceived very differently according to the value system of the actor considered or of coefficients that form equivalences among various units (such as value of time, value of noise, or value of life).

The analyst's skills and intellectual honesty, then, are of the greatest importance at this level of the modeling effort (see Sections 2.2.5 and 2.2.6). Even so, he must respect a few logical requirements that both guide him and allow him to justify the options he has taken. We shall address this comment more explicitly in a subsequent volume where we define the necessary logical requirements and generalize them to account for thresholds. Without going into detail, we propose the following logical requirements.

Exhaustiveness: The loss of information that inevitably occurs when going from Γ(a) to a family F of n criteria[2] must be carefully monitored to avoid situations in which two actions a and a′ are considered equivalent on each of the n criteria but in which arguments could be made against a situation of indifference when considering Γ(a) and Γ(a′).

Cohesiveness: This requirement deals with the compatibility that must exist between the role that each criterion plays when considering preferences along its specific significance axis and the more comprehensive role that a family F plays when integrating all the consequences into comprehensive preferences. Consider, for example, two actions a and b that are indifferent in the sense of comprehensive preferences, and two other actions a′ and b′, respectively, that result from degrading the performance of a on one criterion and improving the performance of b on some other criterion. The requirement of cohesiveness implies that b′ must outrank a′ with respect to comprehensive preferences.

[1] *This is not a direct corollary of the clarity and universality principles presented in Section 8.2.5 a and b.*

[2] *When F contains criteria other than true criteria, the thresholds that define the other criteria are an integral part of the family and, therefore, of the information it portrays.*

Nonredundancy: None of the n criteria of F is considered redundant if leaving out some criterion would form a family that would no longer satisfy one or both of the preceding requirements. The reader can easily verify that leaving out a criterion of a family F could violate the exhaustiveness requirement. To see that it could also violate the cohesiveness requirement, consider the effect of leaving out the criterion g_3 in the family of criteria presented in Section 10.3.3 dealing with the example of the mayor's preferences.

Any family of criteria satisfying the three preceding requirements is said to be coherent (for more details, see Roy and Bouyssou, 1993, chapter 2). Let us point out that this definition implies no independence conditions other than those required by the requirements of cohesiveness and nonredundancy. We return to this point in Section 10.3.3.

The reader wishing to consider these issues in a specific context can reread, for example, the sections dealing with the "Media-Planning" case, Example 5. We next conclude this example, emphasizing those aspects dealing with the development of a criterion family.

Example 5: Media-planning (from Section 8.2.2.1)

Section 8.1.5 shows how it may be arbitrary to evaluate the performance of each of the potential periodicals (elements of A'_p selected as being relevant to a specific campaign plan p) according to a single criterion, such as the cost per thousand targeted readers. Given the myriad of effects and attributes that could be relevant, one could start to respond to the agency director's concerns by building a fairly complete list of criteria (e.g., one in which none of the elements in Table 8.1.1 is omitted); from this list the director of the study of a plan p could choose the relatively few criteria that seem the most appropriate for building a coherent family F of characteristics of p.

Section 8.2.2.1 suggests that most of the criteria of F should be pseudo-criteria or, perhaps, semi- or pre-criteria. Table 10.1.1, which is based on a case presented in Abgueguen (1971), gives an example of one such family. The numbers in such a table cannot appear arbitrary if the table is to be a useful element of a flexible procedure that guides the development and comparison (conforming to the desiderata of Section 3.3) of combinations of titles forming campaign plans. In particular, each criterion's significance axis must be made explicit and related to factual elements. This allows the support to be defined in such a way that the assessment of a periodical's performance on the relevant scale can be justified within the limits of the thresholds used. As such, one would have to be specific about the way in which the editorial and the advertising contexts were evaluated and how these evaluations were combined in the "context" criterion. It is also necessary to be explicit as to how the cost per thousand targeted readers was calculated.

As indicated in Section 6.1.3, the idea of attractive plans can be based on a ranking of the titles of A'_p that takes into account all the criteria. It was exactly this type of media-planning problem that led to the ranking procedure called ELECTRE II (see Roy and Bertier, 1973). A description of this method and its application to this example can be found in de Montgolfier and Bertier (1978) and will be presented in the second volume.

Table 10.1.1: Example of a coherent criterion family and the associated performance tableau for potential use in guiding the conception of an advertising campaign

	Criteria					
	Context	Cost per 1000 (in Francs)	Regu-larity	Strength	Corre-spon-dence	Pres-tige
Indifference Thresholds	1	15 %	1	1	1	1
Preference Thresholds	2	25 %	2	2	2	2
Supports						
L'Express	4	114	6	3	9	7
Jours de France	10	58	5	6	7	9
Modes de Paris	7	48	7	5	5	5
Mlle Age Tendre	6	77	6	3	5	3
Elle	10	51	4	5	8	9
Femmes d'Aujourd'hui	7	62	6	5	6	5
Intimité	5	74	8	2	5	3
Nous Deux	5	125	7	2	5	3
Modes & Travaux	6	55	6	9	5	4
Echo de la Mode	6	86	6	3	5	4
Marie-Claire	10	59	4	6	8	9
Marie-France	9	59	4	5	7	7
Femmes Pratiques	7	51	4	4	7	6
Jardin des Modes	10	65	4	2	8	10

Except for the second criterion, which is defined on a monetary scale, the other 5 criteria are based on scales with 11 degrees encoded from 0 to 10. The thresholds of these 5 criteria are constant in absolute value. The thresholds for the "cost" criterion are constant in relative value.

(end of Example 5)

10.2 PERFORMANCE TABLEAU

Consider a set A′ of potential actions, where A′ could be equal to A, and n criteria g_1, ..., g_n that form a coherent family F. We denote this as g_j, $j \in F$. Consider a table, such as Table 10.1.1, that contains the values of $g_j(a)$ for all j in F and all a in A′ and, if applicable, the characteristics of the threshold functions $q_j(g_j(a))$ and $p_j(g_j(a))$. We shall call this table the **performance tableau of A′ on F**.

Recall that some of the numbers $g_j(a)$ may have no cardinal interpretation. For example, they might be defined on a purely ordinal scale, such as some of those in Table 7.1.1. This is why we use the term **performance** for a general $g_j(a)$. When it is useful to emphasize the quantitative nature, we shall replace "performance" by "valuation" (when the criterion is a gradation) or by "utility" (when the criterion is a measure). Therefore, Table 7.1.2 can be considered a valuation tableau, where the threshold information has been omitted.

For most problems, the performance tableau provides an interesting summary for the decision maker. In the case of the description problematic P.δ (Section 6.1.4), it usually represents the end of the analysis. Since A′ groups the actions considered most relevant, the significance axes of the criteria are understood and accepted, the operational means of defining the functions used to calculate the various performance levels have been made explicit, and the thresholds clearly indicate the valid domain of the numbers, this tableau is an instrument for invaluable dialogue for the decision maker and the other stakeholders in the decision process.

Let us now look at how some stakeholders have a tendency to use and quite often abuse this tableau. When the criteria are on the same scale (e.g., ratings on a scale of 0 to 10, or a percentage index as in Table 7.1.3), it is often tempting simply to add the performance levels to obtain total scores that can be used to compare the elements of A′. When the scales bring into play heterogeneous units, as in Table 10.1.1, one often normalizes the performance levels in the tableau to make the units homogeneous and then applies the same technique. The two most frequently observed normalization procedures are:

- converting into percentages: for each criterion, a minimum and maximum are chosen and assigned levels of 0 and 100, and the corresponding linear transformation is applied to the initial performance level;
- converting into ranks: for each criterion, the actions are ranked according to increasing performance levels and are assigned a score corresponding to their rank in the resulting order.

To demonstrate this second procedure, consider a comparison of l'Express and l'Echo de la Mode on the basis of the data presented in Table 10.1.1. Since l'Express obtains a score much larger than l'Echo de la Mode – 37 to 23, according to Table 10.2.1 – one may be tempted to conclude that l'Express is to be used in the plan instead of l'Echo de la Mode. Note that if the score had been determined through an index based on the "percentage normalization" principle suggested above, the result would have been:

- for l'Express:

$$(0\times 100)+(\frac{11}{77}\times 100)+(\frac{1}{2}\times 100)+(\frac{1}{7}\times 100)+(1\times 100)+(\frac{4}{7}\times 100)=236;$$

– for l'Echo de la Mode:

$$(\frac{1}{3}\times 100)+(\frac{39}{77}\times 100)+(\frac{1}{2}\times 100)+(\frac{1}{7}\times 100)+(0\times 100)+(\frac{1}{7}\times 100)=163;$$

which would suggest the same conclusion in this case, although it does not always have to be so. When comparing the individual performance levels in the example, however, note that l'Echo de la Mode is better on the first two criteria, l'Express is better on the last two criteria, and the two are the same on the middle two criteria. There is no reason to believe that the two differences that indicate better performance by l'Express (9 ⊖ 5 and 7 ⊖ 4) undeniably outweigh the two differences where l'Echo de la Mode performs better (6 ⊖ 4 and 86 ⊖ 114). We could conclude just the opposite if, for example:

- the differences between 6 and 4 on the "context" criterion and between 4 and 7 on the "prestige" criterion are small enough to be considered negligible;
- the increase in performance from 5 to 9 on the "correspondence" criterion represents only a small advantage, either because the "correspondence" criterion is not considered very important or because scores above 5 or 6 are all considered very good, and the advantage or utility of any unit increase from a score above this level decreases very rapidly;
- the 28-franc increase in cost per thousand targeted readers, on the other hand, is considered to be high and, therefore, not worth the gain in compatibility.

Table 10.2.1: Use of the sum of ranks for comparing two periodicals based on the data of Table 10.1.1

Periodicals to Compare	Context	Cost per 1000 (in Francs)	Regu-larity	Strength	Corre-spon-dence	Pres-tige	Total score according to ranks
L'Express Performance Level	4	114	6	3	9	7	
Rank	1	2	7	4	14	9	37
L'Echo de la Mode Performance Level	6	86	6	3	5	4	
Rank	4	3	7	4	1	4	23

For the second ("cost") criterion only, preferences decrease with increasing criterion value. Ranks for equal performance levels are defined such that, for example, when the two lowest ranked periodicals on a criterion are equivalent on that criterion, the next periodical in the order is given a rank of 3, no matter how may others are equivalent with it.

This numerical example illustrates the strong assumptions that are implicitly assumed to hold when one of the two procedures mentioned above is used and should, therefore, caution the reader against using them indiscriminately. Assigning a weight to each criterion allows the substitution of a weighted sum for the simple addition and avoids treating all the criteria as if they were equally important; it is, therefore, preferable.

Nevertheless, the arithmetic manipulations can sometimes lead to somewhat arbitrary compensations between the advantages and disadvantages. The reader can think about this issue by considering a similar example under the hypotheses that all of the six criteria have exactly the same importance, that differences less than certain thresholds have little or no significance, and that differences of the same magnitude at different places on the same scale can reflect preference differences that are not indifferent.

In spite of the criticisms presented above, the two simple aggregation procedures presented, as well as other related procedures, are not without merit. When one uses them with caution, especially keeping in mind the weaknesses of the compensatory mechanism on which they are based, they can help reduce the size of the problem, highlight important tendencies, or serve as a departure point for discussion.

As we shall see in the next chapter, aggregating performance levels becomes crucial when the decision aid considers the choice, sorting, or ranking problematics. Here, we simply mention that even if the performance tableau does not represent the final goal of the study, it is still usually a necessary construct. Although there may be some cases where this tableau may not be pertinent – e.g., when finding the optimum of a single criterion over an infinite set A – when there is more than one criterion, considering a tableau relative to a suitably defined subset A' of A is in order (see Section 11.3).

10.3 DESCRIPTIVE AND CONSTRUCTIVE APPROACHES: PROBLEM OF CRITERIA DEPENDENCE

Regardless of the problematic, decision makers, analysts, and critical observers will be more comfortable when there seems to be no dependence among the criteria of the analysis. We address this problem of dependence from two perspectives:

1) the meaning that one gives to the idea of dependence; we shall show that there are many definitions and different forms of dependence;
2) the objective reasons for desiring a given form of independence among the n criteria and how reasonable it is to propose such independence.

We address these issues below but note here that we are only interested in the linkages among criteria that could affect the underlying preference model of the decision aid. This model is usually developed from one of two seemingly opposite approaches with respect to dependence. The first subsection presents these two approaches. The following two subsections discuss the extent to which they could influence the very conception of the coherent criteria family.

10.3.1 Descriptive and constructive approaches

In what we call the **descriptive** approach, preference modeling is performed in the framework of the following hypotheses:

1) There exists a "decision maker" in whose mind a value system and logical principles (indicative of a certain form of rationality) interact to predetermine unambiguously how any two actions a and a′ compare with each other. This implies that, "somewhere," there exists a system of preference relations before the study is begun. However, this SPR and even the fundamental concepts upon which it is based are not explicit for the decision maker. They only preexist in a latent state.

2) The task is to capture (without influencing) certain elements of this SPR so that the part which is made explicit can be described, if not completely, at least as exactly as possible. When used, this description must allow a "prediction" of how two actions would compare with each other in the eyes of the decision maker on the basis of a model of the consequences that either uses the set of state and dispersion indicators directly or uses the summary that is provided by a performance tableau.

The SPR to be described is generally assumed to contain no incomparabilities and to form a complete preorder (or perhaps a semi-order or pseudo-order) on A. As we saw in Chapter 7, for the cases of interest this structure can be represented by a value or utility function defined on A. In this descriptive approach, such a function is assumed to preexist. Moreover, if one decides to model the function, either through state indicators and dispersion indices or through the n criteria of a family that synthesize these elements, it is also assumed that the function can be easily used to answer the majority of the questions the decision maker might ask, given that she understands the relevance of the function obtained.

The second approach is what we call the **constructive** approach. In this approach preference modeling is performed under the following hypotheses:

1) Two actors interested in the decision can furnish different, even conflicting judgments when comparing two potential actions.
2) An actor (whether she participates or not in making the decision) can be led to modify her preferences, either because she has no initial opinion or because she accepts the relevance of certain arguments presented during the process.
3) There exist concrete significance axes on the basis of which each actor interested in the decision builds, transforms, and justifies her preferences.
4) The task is to use significance axes familiar to the different actors to conceive of a coherent family of criteria. The criteria should allow the construction of one or several systems of preference relations that could be accepted as bases for guiding the decision process and developing the response elements to the questions raised by the decision maker.

It is important to note that the SPR's that are constructed could contain incomparabilities. In some cases, the incomparibilities may reflect the hesitation of some actor or the conflicts among different actors. In other cases, however, the incomparabilities may be the best way for the analyst to take a position on the comparison of two actions when faced with insufficient knowledge of the consequences of the actions or of the value system that should be used. It is also important to note that saying that one construction is more exact than another or that it is wrong or biased has no meaning in such an approach. These ideas would imply a preexisting reality to which one could compare the model developed. In the constructive approach, the only reality considered is one that is continually moving and is influenced by the study. It is, above all, the operational character of the model that indicates its quality.

We provide more details on these approaches and illustrate them in Roy and Bouyssou, 1993. The reader can also see Bouyssou (1984); Roy and Bouyssou (1986); Roy (1987); Bell *et al.* (1988), and Roy (1993). The preceding discussion is sufficient, however, to address the issues of dependence among the criteria raised at the beginning of this section. In particular, we shall show that certain forms of dependence relative to a coherent family of criteria F are not analyzed in the same way when they are interpreted as resulting from a descriptive approach as when they are interpreted as resulting from a constructive approach. This will, in turn, influence the conception of F.

Two basic types of dependence can be considered, according to whether or not they play a major role in the value system underlying the preferences. They will be discussed in the following two subsections.

10.3.2 Structural or statistical dependence among criteria compoments

Here, we consider those components that enter the support of the criteria: the state and dispersion indicators and the factors associated with their definitions. The actors' value systems do not significantly affect these components. Rather, the dependence results from the structural or statistical relations among the components.

To introduce this type of dependence, consider again the highway location example (see Table 7.1.2). Assume that the coherent criterion family includes, among others, the two following criteria:

$g_1(a)$: difference in costs of location a and some reference location a_0;
$g_2(a)$: average time saved for an individual using location a instead of reference location a_0.

Both criteria are related to a common factor: the length of the highway segment. Therefore, the values $g_1(a)$ and $g_2(a)$ would probably be highly correlated over the set A considered. This assumes, of course, that factors such as the existence of archaeological sites or the type of soil in a potential location and the amount of land that would have to be appropriated for the right-of-way of one location do not create too much heterogeneity among the potential locations. If not, then the criteria g_1 and g_2 depend on the length of the segment, and they cannot be considered independent. This dependence is both structural and statistical.

Imagine that the impact of the other factors mentioned above on the variants of interest is negligible and that the criteria for actions in the set A are highly correlated through this relation with segment length. In a descriptive approach, one could consider replacing g_1 and g_2 by a single criterion g_3, where:

$g_3(a)$: difference in lengths of location a and reference location a_0.

When they do not conflict, as is the case here, at least one criterion can be eliminated.

This is not the case in the constructive approach, however. Criteria g_1 and g_2 belong to F because cost and time savings are significance axes that are familiar to the actors and relevant for thinking through their preferences. It is likely that g_3 would not be such a criterion. Moreover, cost is associated with a category of consequences that mostly concerns those financing the project, whereas time savings concern other actors, the users. That is, the consequences captured by these criteria fit into or come from "different pockets." The fact that they are somewhat correlated over the set A is foreign to the concept of preference modeling. In a constructive approach, then, both criteria should remain in the model.

These types of links can have very different origins. For example, they can come from:

- criteria that result from splitting a dimension (Section 9.2.2.2);
- criteria such as g_1 that are not sub-aggregations (Section 9.2.3);
- criteria calling upon a single endogenous factor or different factors that are statistically dependent (see the above example).

It is important to realize that this type of dependence may be related to elements considered in the set A. It is often possible to think of realistic actions that would decrease or even avoid this type of dependence among the criteria. Actions that include tunnels in the alignments would be such actions in the above example: The increased time savings (g_2) that would result from the shorter sections made possible by the tunnels would now be correlated with increased rather than decreased costs (g_1) due to the expenses associated with the tunnels.

In a descriptive approach, the existence of such relations among two or more criteria that cannot be easily broken by adding realistic actions to the set A is generally the sign of an overly rich family of criteria. We say that it is "overly" rich, since a smaller criterion family could be formed that would offer a description of preferences that is as good as that offered by the original criteria family and that would facilitate the modeling of an SPR. This overly rich family may actually inhibit the description desired. Consider, for example, the case where the description takes the form of a value or utility function $U[g_1(a), ..., g_n(a)]$. Establishing such a function to represent a complete, pre-existing preorder is generally easier if the function has fewer arguments. Therefore, it would generally be useful to substitute a single criterion for two or more that are functionally related in a positive manner.

On the other hand, in a constructive approach, such a relation does not weaken the family F when the interrelated criteria are of different degrees of interest to actors or, more generally, as long as they refer to autonomous points of view. When this is the case, the argument of double or triple counting cannot be used to justify eliminating some criterion in favor of another. F is no longer used to provide the **superstructure of a model designed to consider, as faithfully as possible, a reality of fixed preferences**, but to be, **the key words of a language that could be accepted as a basis to develop, reason, and transform dynamic, conflicting preferences**. Therefore, the arguments presented above are no longer relevant in the framework of a descriptive approach.

10.3.3 Value dependence: Links between significance axis preferences and exterior consequences

In defining a criterion function g (Def. 9.1.1) we referred to the existence of a significance axis on which any two potential actions a and a′ could be compared based only on the aspects of the consequences considered by g. The last part of Definition 9.1.1 implies that this comparison makes sense when disregarding all the aspects of the consequences not modeled in the support of g, that is, that we can define preference relations **restricted to the significance axis of g**. These preferences must, therefore, be the same as those that would result when considering comprehensive preferences (i.e., those based on the consideration of all the consequences) of two actions that are equivalent on all aspects of the consequences not modeled in the support of g. The comparison of two actions a and a′ whose consequences differ only in aspects considered in the support of g, therefore, cannot be influenced by the identical (for a and a′) performance on other criteria of the family. This constitutes a type of independence of each significance axis relative to the set of the other n – 1 axes in the family. We call this type of independence one of **isolatibility**.

This form of independence can influence the choice of significance axes of the criteria in a family F. Consider the example of the mayor's preferences (Section 7.2.1). In this example, only two categories of consequences are considered:

– the number of jobs created by each project;
– the cost of the project to the municipality.

Note that considering only these two categories of consequences already implies an acceptance of reasoning that is based on assuming "all other things being equal."

Assume that the analyst attempts to use two significance axes and arrives at the two following criteria:

$g_1(a)$: number of jobs created by the project, defined in terms of the certainty equivalents based on a utility function;
$g_2(a)$: cost of the project to the municipality, defined on a discrete scale based on a consensus of expert opinion.

If these criteria are to play their intended roles in aiding the decision, the mayor must be willing to compare any two projects a and a′ according to the following hypotheses:

when $g_1(a) = g_1(a')$, a and a′ can be compared only according to their costs, no matter what the common value of g_1;

when $g_2(a) = g_2(a')$, a and a′ can be compared only according to the number of jobs that they create, no matter what the common value of g_2.

The mayor might reject the second hypothesis based on an argument such as: "If Project a guarantees 50 new jobs while Project a′ would give me either 110 or 10 new jobs, each with a 50 percent chance, I might prefer a′ if the common cost of the two projects is small enough. I would certainly prefer a if the cost of the two projects were high, however, since I would not want to take the chance of not being able to recover the costs of the chosen project." Note that this argument contradicts the first hypothesis also.

When faced with such an argument, it becomes difficult to base a descriptive approach on the family $\{g_1, g_2\}$. When it becomes a case of describing a pre-existing SPR, fixing criteria beforehand often causes dependence problems of the form considered here. The descriptive approach must thus be based directly on the state and dispersion indicators. In this example, the dispersion appears in probabilistic terms. The point criterion used is based on utility theory. Under these conditions, the type of dependence that results from the above argument is commonly called **utility dependence** (see Keeney and Raiffa, 1976).

The mayor's argument would have a different effect in a constructive approach: F would not be considered a coherent family of criteria due to a lack of exhaustiveness (Section 10.1). Indeed, one could imagine two projects a and a′ between which the mayor would not be indifferent even though:

$$g_1(a) = g_1(a') \text{ and } g_2(a) = g_2(a').$$

Analyzing the argument points up an important point of view dealing with the significance axis: "risk of not recovering project costs." Let g_3 be a criterion associated with this axis. Each criterion in the family $\{g_1, g_2, g_3\}$ is now **isolatible**; that is, it corresponds to a significance axis that, with respect to the set of the others, allows one to reason based on "all other things being equal." When faced with the argument considered, this is the type of family to which a constructive approach leads.

Thus the existence of utility dependence – i.e., of non-isolatible criteria – in a family F takes on a different meaning and requires different responses depending on the approach that is chosen.

Let us now consider a family F consisting of n criteria, each of which is isolatible with respect to the other n – 1. Let J be a subset of F that contains at least two criteria. We say that J is **preferentially independent in F** if, for any two potential actions a and a′ considered (see Table 10.3.1):

$g_i(a) = g_i(a')$, $\forall$ i $\in$ F\J $\Leftrightarrow$ a and a′ can be compared only on the basis of the criteria in J, no matter what the common values on the criteria not in J.

Table 10.3.1: J = {g_1, g_2} is preferentially independent in F = {g_1, g_2, g_3} if the result when comparing a and a′ does not depend on the value of y on the g_2 criterion scale

Actions \ Criteria	g_1	g_2	g_3
a	x	y	z
a′	x′	y	z′

In many cases, the analyst will consider this type of independence to be satisfied, either by the SPR that he attempts to describe in a descriptive approach, or in order to construct the SPR's that are to be accepted as a basis for reasoning in a constructive approach. There are also cases where reasoning on the basis of "all other things equal" (this time when considering not one, but at least two criteria) cannot be accepted. Consider again the case of the mayor and projects a and a′ characterized, using the notation of Table 10.3.1, by:

[x = 50, z = 0 (no risk)] and [x′ = 55, z′ = 1 (significant risk)].

The mayor may argue that the comparison of a and a′ can depend on whether the cost is low or high. In this case, J = {g_1, g_3} is no longer preferentially independent, or **independent in preferences**, in F.

Note that in this set, the lack of preferential independence stems from the fact that the significance axes considered cannot be reduced to a single axis that would allow a unique criterion to be substituted for the subset J. If J is preferentially independent in F, one could define a criterion g_j whose support would be the union of the supports of the different criteria of J and that would lead to the same comparison of any two actions that are equivalent on the criteria of F\J as the comparison that would be obtained in a pre-existing SPR or in a SPR constructed from the same data. For this reason, we shall say that the criteria of J are **reducible** in F. This property is similar to that of independence in preferences (see Ting, 1971). However, if criterion g_j, which is obtained by such a reduction, is formally equivalent for describing or constructing preferences, its significance axis may seem artificial and, therefore, be inappropriate for a constructive approach.

We address this form of independence further in another volume. Here, we only wish to emphasize that the presence of one or more subsets of preferentially dependent criteria in F makes the modeling effort much more complicated in problematics α, β, and γ when using either a descriptive or constructive approach. Luckily, the repercussions of the lack of preferential independence are usually of a second order, and the analyst can often disregard them. When he cannot do so, he must pay careful attention to the logical faults that they could introduce in the discussions related to the performance tableau in problematic P.δ.

10.4 MOTIVATION FOR MULTIPLE CRITERIA

A busy analyst will frequently try to simplify both evaluation models based on state and dispersion indicators and coherent criterion families by trying to aggregate all the pertinent consequences in a unique criterion. In such a **single criterion analysis**, each potential action is valued on a single significance axis chosen beforehand. This axis may have a relatively concrete meaning: benefit, profitability rate, gain for the community, utility for the decision maker.

Although such a simplification works well in simple problems, as soon as the consequence cloud becomes even slightly more complex, this single criterion analysis will lead to only a minor reduction in effort and, possibly, some personal satisfaction to those with reductionist tendencies. Such an analysis requires that each of the (possibly very heterogeneous) consequences be quantified in the common units of the chosen significance axis, which often leads to excluding those aspects of the consequences that are hard to capture in such a system of representation. This is one consequence of the instrumental bias denounced in Section 2.2.6. Moreover, quantifying these aspects requires a, perhaps implicit, reliance on constructs such as reference prices or equivalence rates that are unique to the decision maker and directly reflect her value system. The same is true for the aggregation logic. Therefore, this single criterion does not usually possess the qualities of intelligibility, acceptability, and exhaustiveness that are required of a coherent family of criteria. As we indicated in Section 10.1, this family must be conceived in such a way as to bring about a consensus and to act as an instrument of communication.

A **multicriteria** analysis differs from a single criterion analysis in that it aims to make explicit a coherent family of criteria (not reduced to a single element at the outset) that will serve as an intelligible, acceptable, and exhaustive instrument of communication allowing conception, justification, and transformation of preferences within the decision process. Obviously, nothing prohibits the n criteria of F from later being aggregated into a single, more complex criterion that integrates information not directly related to the consequences. We shall see in Section 11.1 that this type of information takes on an inter-criteria meaning that reflects the personalities of specific groups of actors or decision makers. It is important to realize that this approach is different from single criterion analysis. In single criterion analysis, aspects that are recognized by all as being related to the consequences are mixed early on with aspects that result when reducing consequences to a common unit by an accounting procedure that is inevitably somewhat arbitrary and influenced by a very personal value system.

When F does not become a unanimous, clear, and complete reference, the thought process can be seriously weakened and the evolution of the decision process can be derailed.

To make the preceding ideas more concrete, we introduce a few notions that will be addressed further in Chapters 11 and 12.

10.4.1 Dominance

The definition of **dominance** is well-known. A binary dominance relation Δ_F (in its broad form) is defined on the set A by saying:

$$a' \, \Delta_F \, a \Leftrightarrow g_i(a') \geq g_i(a), \; \forall \, i \in F. \qquad \text{(r 10.4.1)}$$

First, note that this concept adds no information with respect to the outranking associated with a single criterion. In the multicriteria case, it should be clear that the dominance of a by a′ must reflect an outranking of a by a′ for each actor who accepts Δ_F as a reference. In other words:

$$a' \, \Delta_F \, a \Rightarrow a' \, S \, a. \qquad \text{(r 10.4.2)}$$

Note that this implication holds, whatever the value system of the actor considered. In particular, two actors who do not give the same importance to the various criteria of F will nevertheless agree that a′ is at least as good as a whenever (r 10.4.1) holds. Therefore, dominance is always of interest when $n > 1$.

The reader can easily verify that Δ_F defines a partial preorder structure on A. This leads to the definition of an **efficient action**. An action a^* is said to be efficient if and only if:

$$\text{there exists no action } a \in A \text{ satisfying } a \, \Delta_F \, a^* \text{ and not } a^* \, \Delta_F \, a. \qquad \text{(r 10.4.3)}$$

In other words an action is efficient if it is impossible to find another action in A that has better performance on any one criterion without having worse performance on at least one other criterion. Efficient actions are also often called Pareto optimal actions.

In many problems, only efficient actions need to be considered. Dominance, therefore, allows a screening of the actions of A that can significantly reduce its size. We note that although it may be relatively simple to find the subset A^* of efficient actions when A contains a limited number of actions, this is not so when the number of actions in A is large or infinite, which is generally the case when A is a set of real-valued decision variable vectors[3] (see Section 5.1).

The fact that the definition of dominance only calls upon F and not on more specific information about the value systems of the various actors has some disadvantages:

– the partial preorder defined by Δ_F is weak; that is, assuming that:

$$\text{not } a' \, \Delta_F \, a \text{ and not } a \, \Delta_F \, a' \Rightarrow a' \, R_F \, a, \qquad \text{(r 10.4.4)}$$

there would be many pairs of actions satisfying the incomparability relation R_F;

[3] *See Goicoechea et al. (1982) or Chankong and Haimes (1983).*

– except for those cases where all the n criteria are true criteria, supplementary information on the value system or systems is necessary to separate the cases in which the dominance reflects indifference from those in which it reflects strict preference: Specifically, the presence of thresholds prohibits claiming that if one of the inequalities of (r 10.4.1) is strict, then indifference must be excluded; note that this demonstrates the difficulty of enriching the dominance concept by considering thresholds.

10.4.2 Rates of substitution

When $n > 1$ in a δ-problematic, one is tempted to try to enrich the weak dominance relation for pragmatic reasons. Such reasons lead to replacing certain situations of incomparability (in the sense of R_F (r 10.4.4)) by situations of indifference or of preference. As when comparing l'Echo de la Mode and l'Express (Section 10.2, Ex. 5) this often results in "trade-offs" between the "positive differences" and "negative differences" that would be derived from two lines of the performance tableau corresponding to actions a′ and a, where certain criteria are in favor of a′ while others are in favor of a. The concept of a rate of substitution is one that can be used to clarify this idea of compensation. It expresses the minimal gain on a criterion that would be necessary to compensate the loss of a "unit" on another criterion. The following gives a more precise description of the notion in the general case of a pseudo-criteria.

Let a be an action characterized by performance measures $g_1(a), \ldots, g_n(a) = \underline{g}$. Assume that the performance level of g is changed by degrading measure g_j from $g_j(a)$ to $g_j(a) - 1$, where the chosen unit is small, but significant with respect to the thresholds – for example, it could be equal to the preference threshold. The **rate of substitution** in $\underline{g}$ of criterion i with respect to criterion j is the minimal increase $r_{ij}(\underline{g})$ necessary in $g_i(a)$ to be able to **compensate** the unit degradation in the value of $g_j(a)$. In this definition, to compensate means that the dummy action a_{ij} characterized by:

$$\begin{aligned} g_k(a_{ij}) &= g_k(a), \text{ for } k \neq i, j \\ g_j(a_{ij}) &= g_j(a) - 1 \\ g_i(a_{ij}) &= g_i(a) + r_{ij}(\underline{g}), \end{aligned} \qquad \text{(r 10.4.5)}$$

is considered indifferent to the action a.

This notion of compensation implicitly refers to an actor Z who judges the smallest value $r_{ij}(\underline{g})$ compatible with the desired indifference. Therefore, we really should speak of a rate of substitution for a specified actor Z. This rate can always be thought of as **the value, expressed in the "currency of criterion i," that Z gives to a unit of criterion j**.

This definition of rate of substitution leads to the following comments:

1) The value of the rate $r_{ij}(\underline{g})$ could depend on:

- the vector $\underline{g}$ forming the basis of the positive and negative differences considered by Z; the influence of the components of vector $\underline{g}$ is usually more pronounced as the value of the criterion considered approaches an extremity of its scale;
- the unit considered on the significance axis of criterion g_j: the existence of thresholds prohibits reasoning on the basis of the limit of the rate's value as the unit considered approaches 0 (which is traditionally done in the case of true criteria).

2) The interpretation of the rate $r_{ij}(\underline{g})$ depends on the type of approach taken:

- in a descriptive approach, the rate is meant to represent a pre-existing reality (in the mind of actor Z) and, therefore, calls up an image of wondering how it varies as a function of a given component of the vector $\underline{g}$ and how good is the approximation of a given hypothesis;
- in a constructive approach, one can no longer refer to "a true value of the rate," and therefore, it becomes much less important to identify the actor Z associated with it; the rate can thus become a tool that allows different actors to confront their opinions, to understand their preferences more clearly, to focus their intentions as to the relative importance that they believe should be attributed to the various criteria.

Unlike dominance, the rate of substitution is strictly dependent on the value system considered. Since single criterion analyses immediately aggregate the two categories of consequences considered by criteria i and j, they integrate the more or less explicit information about the value of the various rates $r_{ij}(\underline{g})$ into the model at a very early stage. When this information is not known (descriptive approach) or poorly chosen or negotiated (constructive approach), this type of analysis will lead to some confusion or obscurity in that phase of preference modeling. The ability to use this information effectively in the process and, therefore, the quality of the decision aid will suffer.

Whatever the approach, determining upper and lower bounds on the rates $r_{ij}(\underline{g})$ is often useful. These bounds represent information on which the various stakeholders in the decision process can agree and that will allow comparisons of actions based upon the union of all of the criteria when dominance does not hold. The study upon which the "highway location" example is based took this direction (for more details, see Roy, 1974; de Montgolfier and Bertier, 1978). Trying to systematize this type of approach, however, goes well beyond the limits of problematic P.δ.

10.4.3 Concordance

When using rates of substitution, it is often assumed (either implicitly, or in an attempt to simplify the problem) that a gain of $k \times r_{ij}(\underline{g})$ in criterion i is required to compensate a loss of k units in criterion j. However, this assumption of proportionality is usually very questionable when k is large and can lead to an action a with very unbalanced performance (i.e., not very good on some criteria and excellent on others) to appear better than an action a′ with average performance on each criterion. When one wants to give priority to actions whose performance levels are as good as possible but relatively well balanced, compensatory ideas should be limited to cases where the differences are relatively small. For larger differences, the notion of concordance is useful.

A criterion i is said to be concordant with the proposition a′ outranks a if comparing values $g_i(a')$ and $g_i(a)$ justifies a′ S_i a.[4]

Let C(a′, a) denote the subset of the criteria of F that are concordant with the proposition a′ S a. Note that the existence of thresholds can allow a criterion i to be in C(a′, a) even though $g_i(a') < g_i(a)$.

To indicate whether a′ prevails over a, one can consider the subsets C(a′, a) and C(a, a′), each of which can be considered a coalition of criteria, one being favorable to a′ S a, the other to a S a′. In a purely noncompensatory perspective, the sizes of the differences have no impact; only the list of the appropriate criteria are of interest. It is no longer a question of compensation weighted by the size of the differences, but of a breakdown of the votes, of the relations of the intrinsic forces to the chosen criteria.

One cannot think about the relative influence of the sets C(a′, a) and C(a, a′) without considering some value system, since the influence is related to the importance given to each of the two sub-families of criteria that are perceived as coalitions. This importance might be quantified, for example, by assigning a constant, called an **importance index**, to each criterion and calculating the importance index for a coalition of criteria C(a′, a) by simply adding the importance indices of the criteria it contains; or some more sophisticated procedure could be devised. The resulting values for C(a′, a) and C(a, a′) might help one to take a position when comparing a′ and a.

10.4.4 Discordance and veto

The arguments for comparing a′ and a based only on the importance of the coalitions C(a′, a) and C(a, a′) might be too simplistic in many cases; it could also be useful to consider what happens with the discordant criteria.

By definition, **a criterion is said to be discordant with the proposition a′ outranks a if it is not concordant with this proposition, and similarly for relations other than outranking.**

To conclude that a′ is at least as good as a based on the argument that C(a′, a) contains criteria that are more important than those in C(a, a′) ignores the fact that action a could be infinitely better than a′ on at least one of the criteria in C(a, a′). Such a discordant criterion conjures up notions of an "oppressed minority." It is advisable to investigate whether the intensity of this discordance and, thus, of the force with which it opposes the proposition a′ S a is sufficient to reject the proposition. This leads to the idea of

---

[4] *In some cases it may be useful to define concordance based on relations other than outranking, primarily on preference or strict preference.*

veto thresholds that can be assigned to certain criteria. Such thresholds indicate a type of limit beyond which the discordance cannot go and allow an outranking.

This idea of a veto threshold must not be confused with that of a minimal level on the scale of criterion i (in absolute terms) for an action to belong to A. The veto threshold pertains not to a level on $g_i(a)$, but to the preference difference between $g_i(a)$ and $g_i(a')$.

In practice the ideas presented above prove to be very fertile for allowing an efficient reading of the performance level tableau and clarifying the decision without requiring a complex model. Their real interest lies in cases with more than two criteria, and they are irrelevant for cases with a single criterion. To use the ideas in a more systematic, formal, and rigorous fashion to construct a model making comprehensive preferences as explicit as possible adds certain complexities. These complexities are only justified when the results help to form the basis:

- either for the personal convictions of a decision maker;
- or for the desirable progress of the decision process.

This will only be so if the decision maker or the stakeholders in the decision process are willing to cooperate in specifying certain intercriterion information closely linked to the value system, such as information related to the relative importance of the criteria or to a somewhat compensatory aggregation rule. The analyst must then enter into a more complex problematic than that of description. This is the subject of the last two chapters of this book.

Chapter 11

MODELING COMPREHENSIVE PREFERENCES: THREE OPERATIONAL APPROACHES FOR PROGRESSING BEYOND THE DESCRIPTION PROBLEMATIC

SUMMARY

Comprehensive preferences consider all consequences relevant to the decision aiding study. The simplest comprehensive preference model consists of an SPR that includes only dominance and incomparability. Problematics other than P.δ require more than this very disaggregate model, however. In Section 11.1.1 we formulate the **performance aggregation problem**. The entire chapter is an attempt to put some structure on the numerous efforts of theoreticians and practitioners to address this problem.

Any attempt to aggregate performance levels requires the analyst to take both formal and informational positions. In his formal position, he will have to consider things such as the types of preference relations compatible with the model, the aggregation logic to be used, and the functional representations of the different criteria. In his informational position, he will have to consider the nature of the intercriterion information required, how this information will be obtained, and procedures to indicate the validity of the information obtained. In Section 11.1.2, we define the **operational approach** as the set of these two types of positions. For the most part, the operational approaches that we present arise directly from one of the three categories defined in Sections 11.2, 11.3, and 11.4. Others appear as *ad hoc* combinations of two of these categories.

Section 11.2 deals with the approach that uses a single criterion to synthesize the preference information without allowing incomparability. This first operational approach (OA1) is based on using an SPR of the form (I, P) with a complete preorder structure or possibly an SPR of the form (I, P, Q) with a pseudo-order structure. This solution to the aggregation problem allows the functional representation g(a) = $V[g_1(a)..., g_n(a)]$. We illustrate the representation V, which we call the **aggregation function**, through the continuation of Example 3. In Section 11.2.2, we discuss the principal types of aggregation functions – weighted sum, additive, multiplicative, lexicographic – and note that V can be defined without an explicit analytical form. The two fundamental positions that characterize OA1 are: i) a position that does not allow incomparability; ii) a position that explicitly states a rule (the aggregation function) addressing the aggregation problem in a synthesizing, exhaustive, and definitive fashion.

Section 11.3 deals with an outranking approach to synthesize preference information. This second operational approach (OA2) is based on making explicit the conditions that characterize soundly established outrankings. This approach leads to an SPR of the form (S, R), with Δ_F being contained in S. Using this SPR to answer the questions posed by the decision maker is not as straightforward as it is with OA1. It is usually necessary to adapt some procedure to the problematic at hand. Instead of an aggregation rule V, this approach leads to a set of tests T presented in (r 11.3.1) which use the conditions that must be verified for the outranking. In ELECTRE methods T uses the concepts of concordance and discordance. We illustrate the ELECTRE I method in the continuation of Example 1.

Approach OA2 is generally associated with a constructive approach and requires a robustness study of the conclusions in light of the arbitrary nature of the intercriterion information. The two fundamental positions that characterize this approach are: i) a position that accepts incomparability; ii) a position that explicitly states a rule or outranking test addressing the aggregation problem in a synthesizing, exhaustive, and definitive fashion.

Section 11.4 deals with the third operational approach (OA3) to the performance aggregation problem. Unlike the other two approaches, OA3 does not make explicit any rules to address the problem in a synthesizing, exhaustive, and definitive fashion. Rather, it is based on an interactive protocol that regulates how the different series of dialogue and processing stages are linked together to develop a solution from local judgments. The manner in which these judgments are put together to lead toward a solution is primarily based on trial and error and is similar to what would come naturally in most everyday decisions (e.g., the family car example). Still, to provide a true decision aid in more complex situations, the analyst will need some protocol that can efficiently organize the successive interactions.

In Section 11.4.2, we describe the interactive protocol phases: explanation, questioning, and processing phases. We discuss the stopping conditions of the OA3 procedures in Section 11.4.3. In a constructive approach, the procedure stops when the questioner or the questionee considers the goal to be achieved or when one of these two parties decides to stop the process. In a descriptive approach, the procedure must converge before stopping. The two fundamental positions that characterize this third approach are: i) a position that gives primary importance to local judgments dealing with a very small number of actions without considering any explicit rule attempting to aggregate, even partially or temporarily, the performance levels; ii) a position that explicitly states a protocol organizing the interaction between the questionee (the decision maker or some actor in the decision process) and the questioner (the analyst or a computer) so as to allow the recommendation to emerge for the problematic considered.

11.1 OPERATIONAL APPROACH AND THE AGGREGATION PROBLEM

Until now, the preference modeling efforts that we have discussed were concerned with either the individual significance axes of the criteria forming a coherent family F or the case of dominance. These types of efforts can be considered disaggregate modeling. As seen in the previous chapter, this disaggregate modeling can be very helpful in reasoning and focussing the discussion when comparing two actions based on their values in performance tableaus. Problematics other than P.δ require a second, more aggregate level of modeling, however, as addressed in this section.

As emphasized at the end of Chapter 10, this more aggregate modeling will require the construction of additional information. We say that this information is constructed rather than collected, since the information is modelled to respond to specific questions in well-defined settings. How the information is constructed depends on how far the analyst wishes to push the comprehensive preference model, specifically, on the degree to which he wishes the model to acknowledge or obscure the incomparabilities R_F of (r 10.4.4). The possible attitudes of the analyst toward this issue led in 1970 to the idea of an operational approach and to an observation that most operational approaches seem to fit in one of three categories (see Roy, 1971; Vincke, 1992; Roy and Bouyssou, 1993). We describe this idea of an operational approach in this section, and then devote Sections 11.2, 11.3, and 11.4, respectively, to the three categories of operational approaches.

11.1.1 The performance aggregation problem

Restricted preferences are preferences that are limited to one of several consequence dimensions or to the significance axis of one of several criteria (see Section 9.1.2,f). Comprehensive preferences, on the other hand, are those that consider all the consequences relevant to the decision aiding study. A comprehensive preference model is a formal representation of comprehensive preferences relative to a set A of potential actions that the analyst believes to be suitable to the decision aiding problem considered (see Chapter 6). Axiom 7.1.1 of Section 7.1 implies that this model has the form of a system of preference relations (SPR) defined on A. In relations (r 10.4.1), (r 10.4.2), and (r 10.4.4), we introduced a particularly simple comprehensive preference model consisting of an SPR of the form (Δ_F, R_F), in which Δ_F (dominance) is a very weak form of outranking common to all value systems of interest. Except in very special cases, however, this model cannot be used to develop a formal selection, assignment, or ordering procedure. A stronger model of comprehensive preferences must be able to:

- compare actions that would be incomparable if only the dominance relation were used for comprehensive comparisons;
- determine dominance distinctions, for example, in terms of indifference or strict preference.

This requires a certain position on how to combine the different performance levels to form a richer comprehensive preference model than (Δ_F, R_F) from the set of disaggregate preferences relative to the significance axes of the various criteria. At this level of the modeling effort, the analyst is confronted with what we call the **performance aggregation problem**. The performance aggregation problem can be presented through the following questions (for more details, see Vincke, 1992; Roy and Bouyssou, 1993, chapter 3):

What intercriterion information and rules are appropriate for associating one and only one of the basic or consolidated preference systems (see Tables 7.1.4 and 7.1.5) with a performance vector couple (g(a), g(a′))? Specifically, when considering elements such as thresholds or available intercriterion information, when is it justified to:

- use an outranking or an indifference relation when there is no dominance;
- transform a dominance into strict preference, weak preference, or indifference;
- construct more complete or specific intercriterion information designed to reduce the number of incomparabilities and better specify the characteristics of the situations involving dominance?

11.1.2 Operational approach and options

To address the performance aggregation problem, the analyst must overcome two types of difficulties. These difficulties arise from the fact that the following two questions usually have complex, confusing, and unclear answers. From the perspective of which actor or decision maker should the performance aggregation be considered? How much ambiguity or incomparability should the SPR allow when there exist differences of opinion or preferences that could change in the future?

To overcome these types of difficulties and build an operational model that aggregates the performance levels, the analyst must take two types of positions:

- a *formal type*, in which he must consider such things as the types of preference relations that are compatible with the model, the aggregation logic to be used, and the functional representations of the different criteria;
- an *informational type*, in which he must consider such things as the nature of the required intercriterion information, how this information will be obtained, and procedures to indicate its validity.

The two types of positions taken jointly constitute what we call the operational approach.

Many factors can influence these positions – factors such as the personalities of the actors participating in the decision process, what these actors expect from the process, or the degree to which their preference systems are already established. Indeed, the analyst should not lose sight of the fact that whether the decision maker is an individual, entity, or community (Section 2.2), her comprehensive preferences will usually emerge only when forced to confront the conflicts among the criteria. Complex fluctuations will very often arise both in the mind of a single decision maker and among the different individuals comprising the entity or community. The relative importance placed on the various criteria can vary greatly from one individual to another or even for one individual during the course of the decision process. The process of thinking through the impacts of the favorable and unfavorable criteria is what will generally lead to preferences.

The vague notion of relative importance of the different criteria is central to the question of aggregation. Nevertheless, as should become clear after the following sections, it cannot be given specific meaning and, therefore, quantified without first defining the aggregation logic (see Vansnick, 1984; McCord *et al.*, 1993; Mousseau, 1995; Roy and Mousseau, 1996). Yet, this logic is rarely fixed ahead of time. As we shall see, it will vary with the circumstances and the individuals involved. The logic may or may not be compensatory; it may or may not exploit the idea of a veto; it may handle preference dependence in different ways; ...

The analyst, faced with the questions raised at the beginning of this subsection, may feel fairly free to determine the architecture of this comprehensive preference model. The phenomena that he is trying to capture are usually complex and, therefore, do not lend themselves to a set of general rules. The analyst will usually have to justify his positions by considering the context of the specific application. The myriad of options found in an operational approach can be only slightly rationalized from some type of deductive, objective reasoning. The intentions of the decision aid make it impossible to dissociate comprehensive preferences from the ways in which the concepts are presented and discussed or from the ways in which the preferences are eventually modeled.[1]

[1] *These ideas are closely related to those presented in Section 2.2.5 relating to analyst neutrality.*

Let us finally emphasize that when determining his operational approach, the analyst must especially consider:

- whether he will adopt a constructive or descriptive approach, since as we shall see, this will affect the meaning of certain elements of the aggregation model (see Roy and Bouyssou, 1986);
- how he expects to operate within the decision process, since this will affect the types of intercriterion information that may be used in the model.

Having studied the ways in which practitioners and researchers have addressed the aggregation problem presented at the end of Section 11.1.1, we propose three broad families of operational approaches. Each represents one way of confronting the messy set of options that need to be taken. The three following sections describe and illustrate these families.

11.2 OPERATIONAL APPROACH 1: USE OF A SINGLE SYNTHESIZING CRITERION WITHOUT INCOMPARABILITIES

11.2.1 General presentation

This most traditional approach addresses the aggregation problem by relying on situations of indifference and strict preference and, in some instances, weak preference. It does not, however, allow any incomparabilities. It ensures the necessary transitivities so that the SPR defined on A has the structure of a complete order or, when weak preferences are allowed, a complete pseudo-order. In realistic problems (Section 7.2.2), such an SPR leads to a functional representation g that becomes a criterion defined on A. The criterion g can also be formed directly from the criteria g_1, ..., g_n that constitute F and be used to define the SPR. (The preorder case corresponds to g being a true criterion; the pseudo-order case corresponds to g being a pseudo-criterion.) Whether g is derived from an SPR or whether it is used to define this SPR, the comprehensive preference model can be characterized by unique synthesizing criterion g:

$$g(a) = V[g_1(a), ..., g_n(a)], \qquad \text{(r 11.2.1)}$$

where V is defined according to some logic and to the intercriterion information. The synthesizing criterion is unique, since only this criterion can replace the n criteria of F. Even so, the criterion can only be defined to an increasing monotonic transformation, since it is not necessarily a gradation or a measure. The intercriterion information used to define V is usually based on rates of substitution (Section 10.4.2), especially in those cases where V has an explicit analytical form. In this type of approach, the analyst's task is to specify this function V, which we call an **aggregation function**.

The structure of the comprehensive preference model underlying this first operational approach dictates the path that the analyst must follow in the decision aiding effort.

- In the case of P.α, he must select the potential actions that lead to a maximum, or near maximum, value of g and see whether this position at the top of the order is robust when considering the imprecision, uncertainty, and inaccurate determination that arise, especially in the construction of the aggregation function V.

- In the case of P.β, he must determine how to associate with each of the predefined categories an interval of values of the function g that determines the limits for assigning actions into the category.

- In the case of P.γ, he must know how to interpret the classes of actions defined by the criterion g, considering that some minor differences in performance levels should be treated as negligible; further, the analyst must especially try to determine the degree to which the order is robust when considering the imprecision, uncertainty, and inaccurate determination that may be present.

We present the following example to illustrate this first operational approach in the context of problematic P.α. We then discuss the common forms of the aggregation function V. The continuation of Example 6[2] illustrates the case of a problematic P.γ. For an application to problematic P.β, we refer the reader to an article (Zollinger, 1982) that presents the loan approval problem in this form. Many other examples are referenced in Roy and Bouyssou (1993, "Bibliographie commentée de cas d'application").

Example 3: Agricultural Development (from Section 8.2.3.2)

Recall that the objective of this problem is the selection of a "best" agricultural development plan $\underline{x}$ from all of those in a set A that is defined as a polyhedron in $\mathbb{R}^m$ (see Section 5.1.2). Assume that the analyst has considered the characteristics of the country and defined the coherent family F based on the following three criteria:[3]

$g_1(\underline{x})$ = the size of the deficit in the balance of trade for agricultural products that would result form plan $\underline{x}$, expressed in millions of dollars per year;
$g_2(\underline{x})$ = the number of people employed in rural areas under plan $\underline{x}$, expressed in thousands;
$g_3(\underline{x})$ = the percentage of the country's calories consumed from its own agricultural products under plan $\underline{x}$;

where each criterion is a linear function of the variables x_1, ..., x_m that constitute the plan $\underline{x}$.

---

[2] *translator's note: Example 6 is continued on Pages 347-353 in the original, French version.*

[3] *The point state indicators defined in Section 8.1.5 could also be used as indicators here.*

If there existed a single plan in A that optimized each of these three functions, it would be the obvious choice for the best plan, up to the limits that the imprecision, uncertainty, and inaccurate determination were taken into account in the definitions of the criteria and the constraints that determined A. But these criteria might very well be in conflict, and to optimize one would probably cause another to be suboptimal. In such a problem, it is often helpful to try to identify efficient plans (see (r 10.4.3) in Section 10.4.1). But since there would be an infinite number of efficient plans,[4] the benefit of doing so is limited.

Suppose that a desire for simplicity, a lack of time, or insufficient resources lead the analyst to opt for OA1. It would be highly unlikely that the unique synthesizing criterion that he must construct would be compatible with a descriptive approach. Even in a constructive approach, this means of addressing the performance aggregation problem requires a simplistic position concerning the two questions presented at the beginning of Section 11.1.2. Even so, this approach can make an important contribution to the decision aid.

Since the units of the three performance indicators are extremely heterogeneous and since the values that would be given to many of the coefficients required to calculate the performance levels would be somewhat arbitrary, the definition of even such a general aggregation function becomes difficult. Moreover, as we emphasized when discussing this example at the beginning of Section 6.1.1, an exact optimum of a function $g = V[g_1, g_2, g_3]$ would probably be of less interest than the accompanying sensitivity analyses and detailed investigations in the neighborhood of the solution. Given the limited budget and amount of time, such sensitivity analyses would only be possible if g was itself a linear function of the variables $x_1, ..., x_m$. These types of arguments lead the analyst to choose:

$$g(\underline{x}) = \lambda_1\, g_1(\underline{x}) + \lambda_2\, g_2(\underline{x}) + \lambda_3\, g_3(\underline{x}).$$

Since dividing $g(\underline{x})$ by a constant would be a monotonic transformation, this form of aggregation relies on only two independent parameters, λ_2/λ_1 and λ_3/λ_1, which can be interpreted as substitution rates r_{12} and r_{13}, respectively. These rates are assumed to be constant, i.e., independent of g in (r 10.4.5). One might think that it would make more sense to allow these rates to vary with the performance levels achieved on each of the criteria. Doing so would greatly complicate the function V, however, and it might be preferable to use a sensitivity analysis to address the concerns associated with constant substitution rates.

The parameters λ_2/λ_1 and λ_3/λ_1, can be interpreted as the price in millions of dollars of a thousand agricultural jobs and of one percentage point on the scale representing independence in the food supply. Based on these interpretations, the analyst can develop

[4] *Only a finite number of efficient plans correspond to the vertices of the polyhedron, but there are still a very large number of vertices representing efficient plans, and there is no reason to eliminate efficient plans not corresponding to vertices without making further assumptions.*

a series of questions for the various government officials of the country that would help him determine a reasonable range for each of these two parameters. Under these conditions, the only difficulties that remain when treating the problem in the fashion alluded to at the beginning of Section 6.1.1 will be technical ones, related to choosing an appropriate linear programming algorithm (especially for the use in post-optimization analyses)[5] and deciding how to conduct a systematic sensitivity analysis that will illustrate the impacts of the different factors. The use of thresholds (Section 9.3.4) can be useful in guiding this analysis and especially in interpreting the results.

To conclude this example, we point out that OA3, which is discussed in Section 11.4, would have also been a suitable approach, as indicated by various applications to similar problems (for example, see Wallenius, 1975; Wallenius *et al.*, 1978; Despontin, 1981, 1982).

(end of Example 3)

11.2.2 Typical aggregation functions

In practice, as in the preceding example, but often for other reasons,[6] a weighted sum is frequently used as the form of the synthesizing criterion:

$$g(a) = \sum_{i=1}^{n} k_i \cdot g_i(a), \; k_i > 0.$$

The coefficients k_i are often called weights in this form of aggregation, and one can always assume $\sum_{i=1}^{n} k_i = 1$. The weights imply constant substitution rates, which can easily be found (see Example 3 above) as:

$$r_{ij}(\underline{g}) = \frac{k_j}{k_i}.$$

The weighted sum aggregation is the only means of aggregation in which the rates $r_{ij}(\underline{g})$ are all independent of the reference vector $\underline{g}$ used to define them. Therefore, this means of aggregation makes fairly strong assumptions of the value system.

Another means of aggregation, called the additive aggregation, generalizes the weighted sum. It results from assuming:

---

[5] *In the Section 5.1.2 references to Fayette, one can find two types of post-optimisation studies related to the Republic of Korea case.*

[6] *Interesting examples of a skillful use of this form of aggregation in the context of OA3 can be found in Zionts (1979, 1981), Belton and Vickers (1990).*

$$\underline{g}(a) = \sum_{i=1}^{n} k_i \cdot v_i[g_i(a)],\ k_i > 0$$

where $v_i[g_i]$ is a nondecreasing monotonic function of g_i, and one can assume that

$$0 \le v_i[g_i] \le 1 \text{ and } \sum_{i=1}^{n} k_i = 1.$$

In the most general additive form, the $v_i[g_i]$ functions can be considered new criteria that are better suited than the initial g_i's for being aggregated by a weighted sum. (We are assuming that at least one $v_i[g_i]$ cannot be obtained by an affine transformation of the initial criterion g_i.) Transforming the criteria from g_i to $v_i[g_i]$ may be necessary, since aggregating by a weighted sum only makes sense if each criterion is a gradation (Def. 9.4.1).[7]

For reasons that can be found elsewhere (e.g., Keeney and Raiffa, 1976; Vincke, 1992; Roy and Bouyssou, 1993, chapter 4), the aggregation formula frequently used in multi-attribute utility theory is the multiplicative aggregation formula:

$$g(a) = \frac{1}{k}\left[\prod_{i=1}^{n} (1 + k \cdot k_i\ g_i(a)) - 1\right],$$

$$\text{with } 0 \le k_i \le 1,\ \sum_{i=1}^{n} k_i \ne 1,\ k > -1,\ k \ne 0.$$

Since a monotonic transformation of the true criteria g(a) representing an SPR leads to the same SPR, if an SPR can be represented by the true criteria g(a) defined by this multiplicative form, it can also be represented by $1 + k \cdot g(a)$ and by:

$$\log(1 + k \cdot g(a)) = \sum_{i=1}^{n} \log(1 + k \cdot k_i g_i(a)).$$

This latter form is that of an additive aggregation.

One other means of aggregation that deserves mention here is a lexicographic aggregation. This consists of ordering the criteria and considering that the first criterion in the order on which the two actions have different performance levels determines the preference order over the two actions. When the scale of each criterion is discrete, this means of aggregation can be shown to be a special case of that of aggregating by a weighted sum. This form of aggregation does not easily account for thresholds when the g_i's are not true criteria, however.

---

[7] *translator's note: At this point, on Pages 345-346, the original French version provides more detail on the concepts of measure and gradation in the context of aggregation by a weighted sum.*

All these aggregation methods imply preference independence in F for any sub-family of criteria. An example of a simple aggregation function that would not imply this independence is:

$$g(a) = [\sum_{i=1}^{n} k_i \cdot g_i(a) + \sum_{\substack{j=1 \\ i \neq j}}^{n} k_{ij} \cdot g_i(a)\ g_j(a)]$$

with $n > 2$, $k_i \geq 0$, $k_{ij} \neq 0$ for at least one i,j pair.

We simply mention here that it is not necessary to have an analytical formula to make the aggregation function V explicit.[8]

Finally, no matter how V is defined, the analyst can always introduce indifference and preference thresholds into the scale of the synthesizing criterion that is finally obtained.

11.2.3 Important comments

The analyst opting for this first operational approach must keep in mind two closely related ideas:

1) Any aggregation into a unique synthesizing criterion implies certain structural properties of the preference system:

   - constant substitution rates in the weighted sum aggregation;
   - preferential independence of any sub-family of F in the aggregation methods presented above;
   - a completely compensatory logic in most of the analytical forms.

   In a constructive approach the analyst must understand the implications of accepting the above as working hypotheses. In a descriptive approach, he must test the degree to which these hypotheses hold. To do so, there must be some means of investigating the hypotheses that would not affect pre-existing preferences. But as suggested by Tversky and Kahneman (1982) and by Gregory *et al.* (1993), whenever the set of consequences is somewhat complex or represents a new way of thinking about a problem – i.e., in real decision aiding applications – preferences will not be stable or well established. In these cases, the questions that would try to elicit these preferences would indeed influence the subject by causing her to discover some aspect or to reinforce or weaken some previously held attraction or tendency.

   Although some claim that any effect of the questions on the preference system is due to the artificial nature of the situations described in the questions, our experience indicates otherwise.

---

[8] *translator's note: Example 6 is continued on Pages 347-353 in the original, French version as an illustration of this point.*

2) Any aggregation into a unique synthesizing criterion implies the collection or construction of intercriterion information. Usually, this information is condensed into numerical values of certain coefficients, for example:

   - weights – i.e., the k_i's in the weighted sum;
   - the ranges and units considered, which will affect the values of the coefficients in the different forms of aggregation;
   - other parameters of the analytical form that define substitution rates or utility functions, as in the multiplicative aggregation.

   To fix the values of these coefficients requires an estimation in a descriptive approach or an exchange of viewpoints that lead to a consensus in a constructive approach. In either case, difficulties may arise in practice.

These two related ideas are above all what must guide the analyst in this modeling task. At this stage he must explicitly address what is to be attributed to the value system underlying the eventual formulation of the decision maker's comprehensive preferences (see Section 2.2.2). These considerations underscore once more the difference between a multicriteria analysis, even when it eventually leads to aggregating the criteria of F into a unique synthesizing criterion, and a single-criterion analysis that avoids the complexity of F by prematurely forming a unique criterion without considering Conditions I and II presented at the beginning of Section 9.2.3.

To conclude, this first operational approach is based on two fundamental positions:

- not allowing any situation of incomparability by adopting an SPR of the form (I, P) or (I, Q, P);
- making explicit a rule (i.e., the aggregation function V) that addresses the performance aggregation problem in a synthesizing, exhaustive, and definitive fashion.

These two positions impose a complete preorder or complete pseudo-order structure on the comprehensive preference system on A. The synthesizing criterion model allows a functional representation of this structure and, therefore, constitutes the most operational form for capturing and exploiting this structure in the context of the decision aiding problematic considered.

11.3 OPERATIONAL APPROACH 2: SYNTHESIS BY OUTRANKING WITH INCOMPARABILITIES

11.3.1 General presentation

Like the first approach, this second operational approach (OA2) relies on an explicit rule to address the performance aggregation problem in a synthesizing, exhaustive, and definitive manner. Unlike OA1, however, OA2 allows situations of incomparability. In fact, OA2 encourages incomparabilities whenever elements such as thresholds, arbitrary intercriterion information, or ignorance of certain aspects of the value system make an outranking relation – and to an even greater extent, an indifference or preference relation – difficult to justify. In this way, it attaches a great deal of importance to explicit conditions that can characterize a soundly established outranking relation. These conditions also allow the identification of indifference or preference situations whenever no ambiguity is present. And, unlike the situation in OA1, such an approach reflects a certain amount of prudence in the analyst by not requiring transitivity of any of the binary relations used. The result is usually an SPR of the form (S, R) and, occasionally, a more refined SPR – for example, one having the form (I, S, R) or (I, P, S, R).

The counterpart of the OA1 aggregation rule V in (r 11.2.1) is a set of conditions that characterize the presence or absence of outrankings (and, perhaps, indifference and strict preference) by taking into account some elements of the performance tableau and intercriterion information. As we shall see in Section 11.3.2, these conditions usually come in the form of tests.

Unlike in OA1, the analyst will not find a clearly dictated path for using the SPR's in OA2 to answer the questions raised in the decision aiding process. The SPR's may lead to situations of dominance, but they will more often lead to situations of outranking that are soundly established. They could also very well entail additional incomparabilities that cannot be resolved at this level of the modeling effort and intransitivities that could only be overcome by much more weakly established outrankings.[9] For these reasons, the OA2 SPR does not exhibit any interesting structural properties that would allow a simple means of:

- selecting $A' \subset A$ in P.α;
- assigning each action in A to one of the categories considered in P.β;
- determining a partial or complete preorder classification of A in P.γ.

The legitimacy of the outrankings established when considering each pair of actions will be what characterizes the SPR. To be able to choose, sort, or rank, the analyst will need to exploit this entire system either by using common sense or by applying a more formal procedure. In the continuation of Example 1 in Section 11.3.2, we sketch out one means

[9] *To investigate transitivity by using, for example, transitive closure consists of considering the outranking a S c as soundly established whenever there exists some b such that a S b and b S c. It is quite possible, however, that comparing the performance levels of a, b, and c according to the outranking tests (Section 11.3.2) would not lead to such a conclusion. One might claim that if comparing the performance levels of a and c directly did not lead to a solidly established a S c, this outranking might still be acceptable in the limit due to the presence of b. We wish to point out the danger of this type of reasoning at this level of the modeling effort, however. For the preference model to be transitive, one must accept the possibility that a S c might be called upon to justify another even more weakly established outranking a S d, even though a S c did not satisfy the condition that would make it a soundly established outranking. This phenomenon of "contagious" propagation – i.e., of accepting poorly as well as soundly justified outrankings on the basis of local considerations – could easily lead to a comprehensive preference model on A that includes outrankings that are too far removed from the types envisioned by the initial conditions.*

of doing so in the context of problematic P.α based on the ELECTRE I method. We illustrate the application to P.β through the loan approval example next.[10] For illustrations in the context of P.γ, we refer the reader to basic articles presenting other ELECTRE methods (see Roy and Bertier, 1973; Roy, 1978; Roy and Hugonnard, 1982). Many other examples are referenced in Roy and Bouyssou (1993, "Bibliographie commentée de cas d'application").

Consider again the loan approval problem (Sections 2.2.4 and 7.1.1). The criterion family F consists of five ratios for which values are automatically calculated based on information contained in any loan application. Assume that these applications arrive each week and that the focus of the decision aid is to implement a procedure that concentrates on those applications that cannot be accepted as "good" or rejected as "bad" with a fairly automatic procedure. The limits of what is considered "definitely acceptable" and "definitely unacceptable" are assumed to have been defined by the "profiles" corresponding to benchmark combinations of these ratios (Table 7.1.3). The actions to be considered by the decision aid, then, appear as those denoted a_i in Table 7.1.3. These actions are not intended to be compared against each other but against the limiting "acceptable" profiles b_1, b_2, ... and the limiting "unacceptable" profiles c_1, c_2, ... of Table 7.1.3.

Under these conditions, the second operational approach consists of defining rules that pay particular attention to the meaning and the precision of the various ratios and that can point out:

- those profiles b_j that are outranked by a_i;
- those profiles b_j that outrank a_i;
- those profiles c_j that are outranked by a_i;
- those profiles c_j that outrank a_i.

One would need to develop a series of steps that would allow the prescribed assignment of any application a_i to one of the three categories envisioned: acceptance without detailed examination, rejection without detailed examination, examination in detail. This tree will, of course, rely upon parameters that characterize Z's attitude towards two categories of risk (see Roy, 1981):

- acceptance without detailed examination of an application that should have been rejected;
- rejection without detailed examination of an application that should have been accepted.

[10] *translator's note: An additional application to P.β is illustrated through the continuation and end of Example 10 in the original, French version of this book.*

11.3.2 Typical outranking tests

In this second operational approach the analyst must develop a set of conditions that test the two performance level vectors $\underline{g}(a')$ and $\underline{g}(a)$ to determine whether a′ S a is soundly established or not. These conditions, which we also call rules, require specific intercriterion information and, in the case of pseudo-criteria, indifference and preference thresholds. The intercriterion information allows outranking relations to hold in cases other than those involving dominance. Unlike in the case of OA1, however, this information and the conditions that integrate it in testing the validity of the outranking are usually conceived in such a way that they are compatible with the joint negation of a′ S a and a S a′, a situation that corresponds to incomparability.

To describe the implications of the statement "a′ S a is soundly established," assume that the analyst has conceived of a certain number of tests to which he can put the performance levels of a′ and a. These tests consist of performing various calculations; let us denote the results by $T[\underline{g}(a'), \underline{g}(a)]$. Comparing these results to certain reference values playing the role of norms allows the analyst to conclude what we shall denote as:

$$T[\underline{g}(a'), \underline{g}(a)] - (\text{yes/no}) \rightarrow a' \; S \; a. \qquad \text{(r 11.3.1)}$$

Let us briefly illustrate in the case of ELECTRE methods. These methods base the outranking on a noncompensatory logic with veto power using the notions of concordance and discordance (for more details, see Bouyssou and Vansnick, 1986; Roy, 1991; Bouyssou, 1992; Roy and Bouyssou, 1993, chapter 5). The tests T consist of checking:

- whether the concordant coalition C(a′,a) (Section 10.4.3) contains enough criteria or enough criteria of sufficient importance;

and:

- that there is no discordant criterion (Section 10.4.4) that leads to an excessive counterperformance of a′ with respect to a.

For example, in **ELECTRE I** (see Roy, 1968; Vincke, 1992; Roy and Bouyssou, 1993, chapter 5), the importance k[C(a′, a)] of a coalition C(a′, a) is defined as the sum of the importance coefficients k_j of the criteria g_j within C(a′, a). A necessary condition of outranking a′ S a is:

$$k[C(a', a)] \geq s \qquad \text{(r 11.3.2)}$$

where s is the concordance threshold chosen by the analyst. A veto threshold v_j (Section 10.4.4) is also introduced for each criterion, and a second necessary condition for a′ S a is:

$$g_j(a') \geq g_j(a) - v_j, \; \forall \, j \in F. \qquad \text{(r 11.3.3)}$$

The outranking statement a′ S a is considered to be soundly established if and only if (r 11.3.2) and (r 11.3.3) hold. The continuation of Example 1 below illustrates this type of rule-based formulation in the case of true criteria. It can also be easily extended to the case of pseudo-criteria (see Roy, 1991; Vincke, 1992).

Because of the many factors of imprecision, uncertainty, and inaccurate determination that affect the consequences and the difficulties that an analyst would have with capturing a value system or common elements of several value systems, there would probably be many [$\underline{g}$(a′), $\underline{g}$(a)] couples for which it would be difficult to say whether a′ S a is in accordance with a well established preference in the eyes of the decision maker or some other set of actors. Therefore, it is often quite natural to think not of one, but of several outranking relations in this operational approach, where the various relations correspond to weaker and weaker levels required for accepting the outranking. This gives rise to richer and richer relations, i.e., relations with fewer and fewer incomparabilities. One can think of indexing each relation by the value of a credibility index d, where d would take values on an ordinal scale that can always be normalized to fall between 0 and 1. In this way, one would define a family of embedded binary relations S^d. The relation could also be considered a binary fuzzy relation, where d corresponds to the membership function (see, e.g., Dubois and Prade, 1980; Perny and Roy, 1992).

To define such a family of binary relations in practice, there are several ways to proceed. For example, one or several parameters used in the definition of the test T could be varied. This technique is used in ELECTRE II, which deals with true criteria and results in two credibility levels of the outranking – a strong and weak level (see Roy and Bertier, 1973). Another way to define the family of binary relations would be to use the concordance and discordance indices to calculate directly the largest value of degree of credibility d such that a′ S^d a holds. This is what is done in ELECTRE III, which deals with pseudo-criteria (see Roy, 1978, 1991). Or, the nature of the test could be structurally varied to make it weaker and weaker. This is what is done in ELECTRE IV, which uses four binary relations dealing with pseudo-criteria with no additional information on their relative importance (see Roy and Hugonnard, 1982).

Many other hypotheses and means of defining the conditions for an outranking test could be imagined within the framework of OA2 (see, especially, Paelinck, 1978; Roubens, 1982; Hinloopen *et al.*, 1983; Brans *et al.*, 1984; Martel *et al.*, 1986; Matarazzo, 1986, 1988; Brans and Mareschal, 1990).

When the outranking model is built, it is necessary to be explicit as to how it will be used for decision aiding. This leads to the concept of operating[11] procedure (see Vanderpooten, 1990; Vincke, 1992).

Example 1: Site Selection (from Section 6.1.1)

This example illustrates the application of OA2 to an α-problematic. From Section 6.1.1, recall that at the end of the first phase of the study, there were only five sites in set A.

a) Performance criteria

The consequences to be considered when reasoning about the future location of the institute can be relatively easily determined from the general context of the problem described in Section 3.1 and from the interviews with the principal actors. Several significant axes appear pertinent for synthesizing all the elementary consequences. The following is a brief summary of their contents.

---

[11] *Also called exploiting procedure in reference to the French terminology: procédure d'exploitation.*

Criterion g_1, Urban Environment: Two elementary consequences are of concern:

- the position of the site with respect to the various city centers;
- the environment (housing conditions, recreational and commercial activities, urban setting, ...) of the area directly surrounding the institute.

Criterion g_2, Industrial Environment: This criterion refers to the suitability of the site to the strategy of involving the institute with the industrial sector.

Criterion g_3, Intellectual Environment: This criterion integrates three elementary consequences related to the institute's accessibility to:

- other institutions of higher education that could complement the scholastic endeavors of the institute;
- a center for continuing education;
- service industries (consulting firms, computing services, ...).

Criterion g_4, Accessibility and Ability to Host Business Travelers: This criterion considers:

- the position of the site with respect to intercity transportation infrastructure (highways, airport, passenger rail terminals);
- the quality of the area's public transportation services (frequency of service, density of access points);
- the capacities of hotels and conference rooms in the area.

Criterion g_5, Implementation Cost: This criterion is influenced primarily by the price of developing a square meter in the area, as well as by the possibilities of sharing certain facilities (sports centers, large amphitheaters) in the area.

Criterion g_6, Implementation Time: The importance of this criterion stems from the fact that the present situation is considered disastrous. Rapid implementation, which is perceived as essential to advancing the institute's proposed strategy, is considered necessary to compete with other similar institutions. It is also believed that delays will make it harder for the institute to adapt to the evolving competitive environment; in fact, there is a feeling that a doubling of time of implementation will more than double the difficulties that the institute faces.

Criterion g_7, Compatibility with Present Characteristics of the Institute: This criterion concerns the "conversion cost" affecting the institute's administrative and instructional personnel, as well as the students and graduates. The client feels that these actors are important, and the six preceding criteria are not sufficient to account for the effect of this aspect on their preferences.

These criteria form a coherent family F according to the requirements of Section 10.1. Unfortunately, the consequences that must be evaluated are mostly qualitative, which makes it difficult to obtain numerical performance levels on their significance axes. Since there are so few sites to consider, however, this difficulty could be overcome by proceeding progressively as follows:

1) A factual description of each of the elementary components mentioned in the criteria descriptions is obtained for each site. These descriptions are based on judgments that are influenced primarily by the sites' locations in the area (see Fig. 6.1.2). Numerical values are obtained only for those consequences (e.g., price, distance, time) that are naturally expressed as such.

2) On the basis of these descriptions, the five sites are then compared successively on criteria g_1 through g_7 and ranked according to decreasing preferences on each criterion (see Table 11.3.1).

3) Each rank for each criterion is assigned a number on a scale between 0 and 10, where 10 is systematically given to the best site on that criterion (see Table 11.3.2). These numbers are determined on the basis of the analysis conducted in 1) above by considering the relative closeness according to the criterion considered of two sites ranked consecutively on that criterion: The difference in values should be larger when the superiority of one site with respect to another is greater. In this way, the reasoning behind the performance level differences on each criterion is consistent with that behind differences in preferences (Section 9.4).

Table 11.3.1: Classification of the sites of Example 1 according to each criterion

Criteria / Actions	g_1 Urban Environment	g_2 Industrial Environment	g_3 Intellectual Environment	g_4 Accessibility and Hospitality	g_5 Implementation Cost	g_6 Implementation Time	g_7 Compatibility
a_1	5th	5th	3rd	5th	1st	1st	4th
a_2	2nd	4th	4th	4th	5th	2nd	2nd
a_3	1st	3rd	2nd	1st	3rd	3rd	1st
a_4	4th	1st	5th	3rd	2nd	4th	5th
a_5	2nd	2nd	1st	2nd	4th	5th	3rd

To illustrate, consider criterion g_2. The distribution of industrial activities anticipated in the metropolitan area places site a_4 in one of the most important industrial zones and locates it at the very geographic center of the different industrial activities of the region. It becomes obvious that a_4 is the best site according to this criterion and, therefore, $g_2(a_4) = 10$. The other sites are then ranked in decreasing order according to their distance from a_4. The superiority of a_4 over a_5 with respect to g_2, which is ranked second on g_2, appears very similar to that of a_5 over a_3, which is ranked third. The superiority of a_4 over a_3, however, does not seem as large as that of a_2 over a_1, the fourth and fifth ranked sites, respectively, on g_2. Reasoning in this way leads to the performance levels on g_2 shown in Table 11.3.2.

The indifferences and preference threshold values also presented in Table 11.3.2 make each criterion a pre-criterion. Proposing $q_j = 0$ and $p_j = 1$, for $j = 1, ..., 7$, might seem rather optimistic, considering the way in which the performance levels were obtained. Nevertheless, we consider the model as a first working hypothesis and investigate the sensitivity of any conclusions to other hypotheses concerning these thresholds.

Table 11.3.2: Performance tableau for the sites of Example 1

	Criteria						
	g_1 Urban Environment	g_2 Industrial Environment	g_3 Intellectual Environment	g_4 Accessibility and Hospitality	g_5 Implementation Cost	g_6 Implementation Time	g_7 Compatibility
q_i	0	0	0	0	0	0	0
p_i	1	1	1	1	1	1	1
Actions							
a_1	2	0	6	1	10	10	3
a_2	8	5	4	3	0	7	8
a_3	10	6	8	10	6	6	10
a_4	4	10	2	5	8	2	2
a_5	8	8	10	8	4	1	5

b) Comprehensive Preferences: Outrankings

From Table 11.3.2, one can see that there is no pair of actions (a_j, a_k), for $j \neq k$ bringing about a situation of dominance. However, note that if a_3 and a_2 could be considered of equal performance on the sixth criterion – i.e., if $g_6(a_3) = g_6(a_2)$ could be accepted – a_3 would dominate a_2, $(a_3\ \Delta_F\ a_2)$. We now use tests (r 11.3.2) and (r 11.3.3) to determine soundly established outrankings.

b1) Concordance

To each criterion j, we must first assign a positive number k_j reflecting the importance that one wishes it to have in this study. These numbers have no relation with substitution rates, and to establish the k_j's one cannot, for example, reason in a compensatory fashion by thinking of multiplying substitution rates times differences in performance levels. The **importance coefficients** k_j must simply:

- be greater for criteria that are judged more important;
- be such that the sum, over all j in C, of the k_j's (which we can denote k[C]) is greater for coalitions C that are judged more important.

To assign values to the k_j's, then, one must consider a ranking of the criteria according to their importance and a comparison of the importance of a few carefully chosen coalitions, which can be considered "super-criteria." A specific method of comparing coalitions can be found in Roy *et al.* (1986) and Mousseau (1995).

The numerical values of the seven coefficients k_j represent a type of intercriterion information: They reflect the values of the actor asked to judge the importance of the criteria and the coalitions. Therefore, to help the director of the institute form an opinion and play her role in the decision process, it would make sense to use not only one set of importance coefficient values reflecting the director's position, but several sets that would account for the different positions held by the other actors. Therefore, four sets of values $\underline{k}^1$, $\underline{k}^2$, $\underline{k}^3$, $\underline{k}^4$ are adopted (see Table 11.3.3). They represent, respectively:

- an equal importance of the seven criteria;
- an equal importance of criteria 1, 2, 3, 4, and 7, and an equal importance of criteria 5 and 6, with the coalition of criteria 5 and 6 having an importance equal to that of the coalition of criteria 1, 2, 3, and 4;
- an equal importance of criteria 3, 4, 5, 6, 7, and an equal importance of criteria 1 and 2, with the coalition of criteria 1 and 2 having an importance equal to that of the coalition of criteria 3 4, 5, and 6;
- an equal importance of criteria 1, 2, 5, and 6, and an equal importance of criteria 3, 4, and 7, with the coalition of criteria 3, 4, and 7 having an importance equal to that of the coalition of criteria 1 and 2.

Table 11.3.3: Values of the importance coefficients of the criteria of Example 1

(For convenience, the values are expressed in multiples of 126, so as to normalize the sum of each of coefficients and to facilitate comparisons among the same criteria in different sets.)

Criteria / Sets of Values	g_1 Urban Environment	g_2 Industrial Environment	g_3 Intellectual Environment	g_4 Accessibility and Hospitality	g_5 Implementation Cost	g_6 Implementation Time	g_7 Compatibility
$\underline{k}(1)$	18/126	18/126	18/126	18/126	18/126	18/126	18/126
$\underline{k}(2)$	14/126	14/126	14/126	14/126	28/126	28/126	14/126
$\underline{k}(3)$	28/126	28/126	14/126	14/126	14/126	14/126	14/126
$\underline{k}(4)$	21/126	21/126	14/126	14/126	21/126	21/126	14/126

The ELECTRE I **concordance index** is the sum of the importance coefficients of the criteria in the concordance coalition. This index can vary from 0, when no criteria belong to the concordance coalition, to 1, when all the criteria belong to the concordance coalition. Table 11.3.4 presents the value of the ELECTRE I concordance index for each pair of actions (a′, a) and each set of importance coefficients k(.).

b2) Discordance

Next, we must assign a veto threshold v_j to each criterion j. This represents a second category of intercriterion information (Section 11.3.3). This veto threshold, which could vary as a function of the interval $[g_j(a'), g_j(a)]$ (see Roy, 1991), defines the set of cases in which the comparison of the performance levels according to criterion j is such that the outranking relation would not be considered soundly established, regardless of the performance levels on the other criteria.

The veto threshold does not have an intrinsic value that can be estimated. Still, the analyst can use real or dummy actions to try to capture, one criterion at a time, the size of the smallest (positive) difference $g_j(a) - g_j(a')$ at which he observes instability in a given actor's perception of preferences or a certain hesitation in the actor's stating that "a′ outranks a" when the differences in favor of a′ on the other criteria would normally lead to the outranking. Usually, the analyst must use some judgment based on rather diffuse opinions to fix the values of these thresholds. This, of course, allows for somewhat arbitrary veto thresholds to be assigned, and the thresholds must later be varied through a reasonable range of values to determine their impact on the eventual choice.

In this example, the veto threshold v_j was considered to be the same regardless of the criterion j considered. It would seem that more important criteria should have lower veto thresholds, implying that the veto thresholds should vary with sets of importance coefficients k(.). However, this level of complexity seems unwarranted for this first-cut analysis. This is especially so when one considers that the values of the veto thresholds are to be varied across a fairly large range to investigate the robustness of the solution.

The **discordance level** D(a′, a) is defined as:[12]

$$D(a', a) = \max_{j \in F} g_j(a) - g_j(a').$$

These discordance levels are presented in Table 11.3.5. It is straightforward to use these numbers, as we do below, to determine the impact of test (r 11.3.3) on the outrankings as a function of the values given to the "parameters" v.

b3) Outranking

The values of two parameters must be fixed to define the cases in which the statement a′ S a is soundly established:

- s: the concordance threshold, the value of which usually falls between 0.5 and $1 - \min_{j \in F} k_j$; ;
- v: the veto threshold, whose value it seems reasonable to restrict between 4 and 8, in this case.

---

[12] *This simple formula is acceptable, since the performance levels were constructed in such a way as to make each criterion a gradation on a common scale. If this had not been the case, the "max" operator could not be used. Introducing thresholds $v_j[g_j(a)]$ that vary with the interval $[g_j(a), g_j(a')]$ avoids the need for criteria that are gradations, but they would still have to be gradable, at least as an approximation, to give meaning to the notion of a discordance level.*

Table 11.3.4: Values of the concordance coefficients for the four sets of importance coefficient values of Table 11.3.3

set $\underline{k}(1)$

a' \ a	a_1	a_2	a_3	a_4	a_5
a_1		0.429	0.286	0.571	0.286
a_2	0.571		0.143	0.571	0.423
a_3	0.714	0.857		0.714	0.714
a_4	0.429	0.429	0.286		0.429
a_5	0.714	0.714	0.286	0.571	

set $\underline{k}(2)$

a' \ a	a_1	a_2	a_3	a_4	a_5
a_1		0.556	0.444	0.667	0.444
a_2	0.444		0.222	0.556	0.444
a_3	0.556	0.778		0.667	0.778
a_4	0.333	0.444	0.333		0.556
a_5	0.556	0.667	0.222	0.444	

set $\underline{k}(3)$

a' \ a	a_1	a_2	a_3	a_4	a_5
a_1		0.333	0.222	0.444	0.222
a_2	0.667		0.111	0.556	0.444
a_3	0.778	0.889		0.667	0.667
a_4	0.556	0.444	0.333		0.444
a_5	0.778	0.778	0.333	0.556	

set $\underline{k}(4)$

a' \ a	a_1	a_2	a_3	a_4	a_5
a_1		0.444	0.333	0.556	0.333
a_2	0.556		0.167	0.556	0.444
a_3	0.667	0.833		0.667	0.722
a_4	0.444	0.444	0.333		0.500
a_5	0.667	0.722	0.278	0.500	

Table 11.3.5: Values of the discordance levels D(a′, a)

a′ \ a	a_1	a_2	a_3	a_4	a_5
a_1		6	9	10	8
a_2	10		7	8	6
a_3	4	1		4	2
a_4	8	6	8		8
a_5	9	6	5	4	

Given the set of values of the importance coefficients $\underline{k}(h)$, the tests T of concordance and non-discordance that provide the necessary and sufficient conditions to assert a′ S a can be written, respectively:

$$k[C(a', a)] \geq s;$$
$$D(a', a) \leq v.$$

Note that applying these tests for s = 1 and any value of v or for any value of s and v = 0 would lead to an outranking that corresponds to dominance – i.e., $S = \Delta_F$. As s decreases or v increases, more incomparabilities R_F will be replaced by outrankings. Therefore, the fixed values of s and v are levels such that any lower or higher values, respectively, would lead to insufficient credibility in the outrankings. Figures 11.3.1 through 11.3.4 use the notation of Figure 7.2.1 to illustrate the results of these tests as a function of the values of s and v.

c) Recommendation: Selected sites

Consider first the set of values $\underline{k}(1)$ presented in Table 11.3.3. The corresponding concordance indices can be found in Table 11.3.4. These indices show that:

- the outranking graph could only contain one arc for s values less than or equal to 0.85 (= 1 – 1/7) and greater than 0.714 (= 1 – 2/7);
- the only critical values of s in the range between 0.75 and 0.55 are 1 – 2/7 = 0.714 and 1 – 3/7 = 0.571. All values of s between 0.71 and 0.58 would lead to the same outranking graph.

Therefore, the effect of the concordance threshold s when using the set of concordance coefficients $\underline{k}(1)$ can be completely portrayed by using the three values 0.55, 0.70, and 0.75. The three corresponding graphs are shown in Figure 11.3.1. The numbers shown above certain arcs indicate the minimum value of the veto threshold v such that the corresponding outranking is compatible with the test of non-discordance. No such number above an arc indicates that the outranking always hold for v between 4 and 8, the values mentioned above as being of interest. Results for the other sets of $\underline{k}(h)$ are shown in a similar fashion in Figures 11.3.2-11.3.4.

Given this first set of importance coefficients, the recommendation to be offered is clear: a_3 should be selected, since it outranks the other four sites for values of s up to 0.70 and for any value of v. It also outranks the other sites for $\underline{k}(3)$ and $\underline{k}(4)$, for values of s up to 0.65. Only in the case of $\underline{k}(2)$, where the coalition of criteria 5 and 6 has the same importance as that of the coalition of criteria 1 through 4, is the conclusion not as clear-cut. In this case, for $s = 0.65$, a_3 remains incomparable with a_1. One might, therefore, recommend the selection of $\{a_1, a_3\}$, since these two sites taken together outrank all the others.

In accord with the client's wishes (see Section 6.1.1), one could investigate the sites to be selected if a_3 were to be rejected by the other actors for some reason. Investigation of the graphs after removing a_3 would tend to indicate a recommendation of a_1 or a_5, which are always incomparable with each other and together almost always outrank a_2 and a_4, at least for $s = 0.55$ and $v = 6$. The only exceptions are in the cases of $\underline{k}(2)$ and $\underline{k}(4)$, where a_4 is incomparable with a_1 and a_5 for $v = 6$ and $s = 0.55$. Even so, if the concordance threshold were decreased to 0.5, a_5 outranks a_4 in $\underline{k}(4)$.

Based on the above, the reader may have the following concerns:

- The type of critical analysis of the outranking graphs illustrated in this problem is only possible when the number of actions is small. What happens when the number of actions increases? The ELECTRE I method was developed to respond to this type of question. The concept of a kernel, which is a classic one in graph theory (see Vanderpooten, 1990), can be used to establish the set A′; the only difficulty is that cycles are possible, as can be seen in Figure 11.3.2, when $s = 0.55$ and $v = 8$.

- How can the definition of an outranking, as given in the tests of (r 11.3.2) and (r 11.3.3), be adapted to account for indifference and preference thresholds such as $q_j = 1$ and $p_j = 2$? The ELECTRE IS method (see Roy, 1991; Vincke, 1992) answers this question. We add that the results that would be obtained in this method when using the thresholds given in Table 11.3.2 do not differ significantly from those presented above.

To finish, we add that a_3 was indeed the site eventually selected for the new location of the institute in the real version of this problem.

Figure 11.3.1: Outrankings for the set of importance coefficients $\underline{k}(1)$ and $0.55 \leq s \leq 0.75$, $4 \leq v \leq 8$

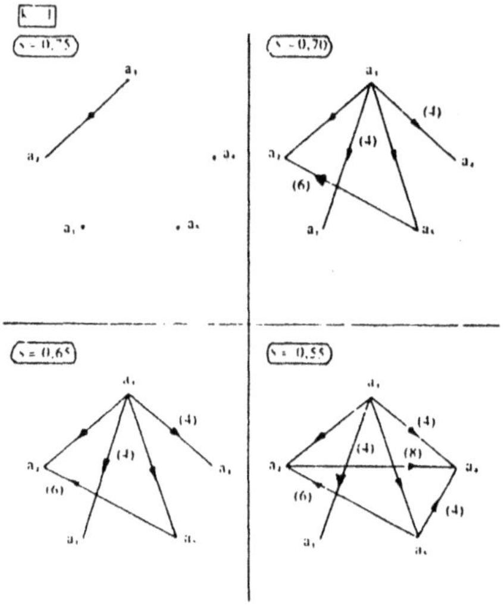

Figure 11.3.2: Outrankings for the set of importance coefficients $\underline{k}(2)$

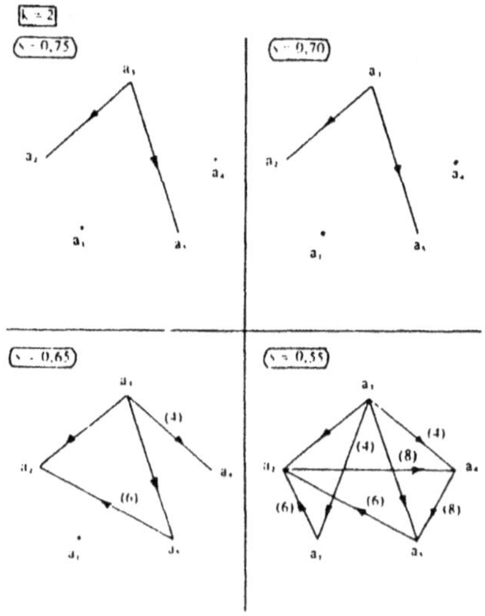

Figure 11.3.3: Outrankings for the set of importance coefficients $\underline{k}(3)$

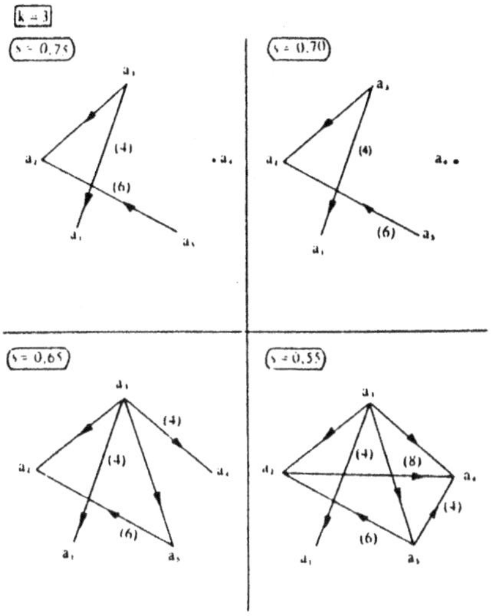

Figure 11.3.4: Outrankings for the set of importance coefficients $\underline{k}(4)$

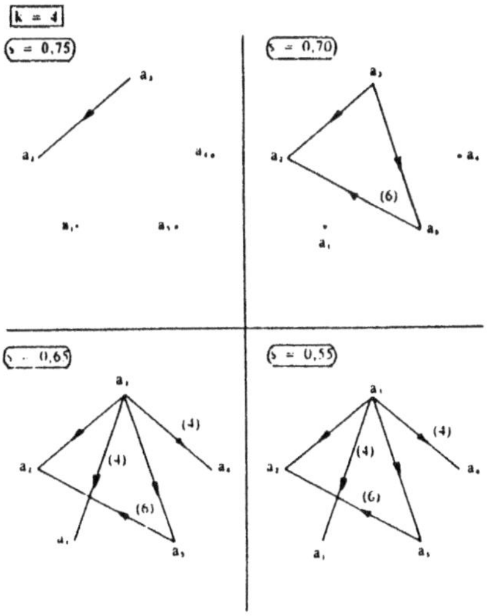

(end of Example 1)

11.3.3 Important comments

The various concepts involved in the test T, such as concordance, discordance, importance of criteria, and veto power; the generally, although not necessarily always, local noncompensatory logic employed; and the numerical values of certain coefficients, such as the importance coefficients of the criteria, and the concordance and veto thresholds, all require the analyst to make certain hypotheses or gather information related to value systems. Therefore, the analyst should again:

- test the validity of these hypotheses and estimate the "true values" of the coefficients of the information in a descriptive approach; when doing so, he will observe that the descriptive approach is rarely compatible with OA2;
- confirm that the hypotheses made and the values assigned to the various coefficients define working hypotheses that are acceptable to the decision maker or the relevant actors in a constructive approach; we note that this is not always as easy as it may seem.

To produce convincing results in this second operational approach, then, will usually require a constructive approach in which the working hypothesis is one in which the values of the principal coefficients are varied from minimum to maximum values considered reasonable. It is not enough to vary one coefficient after the other; the variations must be combined in several possible ways to test the robustness of the major conclusions. Changes in the conclusions with the values of the coefficients could be instructive: They indicate that the decision maker must rationalize her behavior according to some aspect, such as the relative importance of a given criterion or a group of criteria (see Roy and Bouyssou, 1986; Roy *et al.*, 1986).

We wish to point out that the intercriterion information required for a test T in this second approach is generally not as fine as that required to calculate the value of the function V in the first approach. This is primarily because the performance aggregation problem is addressed in OA2 through a system of preference relations that uses incomparabilities in complex, questionable, or conflicting cases and only seeks to establish outrankings that cannot be propagated through transitivity in other cases (see Roy and Mousseau, 1996).

To conclude, this second operational approach OA2 is based on two fundamental positions:

- allowing situations of incomparability and the use of an SPR based on outranking, with several levels of credibility of the outranking, if necessary;
- making explicit a rule (i.e., the outranking test) that addresses the performance aggregation problem in a synthesizing, exhaustive, and definitive fashion.

These two positions impose no notable structure on this SPR "solution." Therefore, there remains a great deal of latitude in using it in the context of the decision aiding problematic considered. (For more details on this point, see the discussion in Roy and Bouyssou, 1993, chapter 6.)

11.4 OPERATIONAL APPROACH 3: INTERACTIVE LOCAL JUDGMENTS WITH TRIAL-AND-ERROR ITERATIONS

11.4.1 General presentation

The fundamental option that was common to the two previous approaches is abandoned in the third operational approach OA3. This approach does not try to make explicit any rule that addresses the performance aggregation problem in a synthesizing, exhaustive, and definitive manner. Rather, it is based on an *ad hoc* sequence of judgments formulated by the decision maker or other actors. These judgments are only meant to have a local meaning, in that they are based either on performance levels in the neighborhood of those of a single action or on a very small number of actions deemed relevant to compare because they are considered neighbors (in the sense of a certain topology given to the problem.)

For these judgments to be emitted, questions must be posed to someone whom we shall call the **questionee**. This can be the decision maker or some other actor. The **questioner** can be either the analyst or a computer. This questioner chooses the types of questions as well as the action or actions that are considered in each question. The approach is interactive, however, in that the responses of the questionee are meant to have a great influence on the order and context of the subsequent questions.

This third operational approach thus consists of establishing a sequence of exchanges, which can be considered a dialogue between the questioner and the questionee (see Vanderpooten, 1989a, 1989b, 1992).

After each question or sequence of exchanges, the answers that are obtained are processed by the questioner in an attempt to use the information contained in the responses to prepare new questions. In this way, this interactive procedure, which is at the heart of OA3, consists of alternating stages of **dialogue** and **processing**. If such a procedure is to be useful in the decision aiding effort, these alternating stages must either lead the decision maker to answer the questions she has posed or allow the analyst to gather the information needed to develop his recommendation. The procedure can lead to:

- the emergence of a small number of actions in A that the decision maker accepts to select, in the case of P.α;
- a determination of the categories to which the decision maker would accept assigning the actions of A, in the case of P.β;
- a way to structure A with a complete or partial order that conforms to the decision maker's preferences, in the case of P.γ.

Such a procedure must, therefore, lead the questionee through a series of exchanges that will allow a choice, a sorting, or a preorder on A. This series, which will lead to the formation of an opinion or even a conviction in the mind of the decision maker and/or the analyst, is certain to include some sort of trial-and-error process. The local judgments voiced during one exchange of the procedure must be reconsidered during later exchanges. The need to reconsider the local judgments arises because during the process the questionee develops a better understanding of the actions to be considered in A and of their performance levels. Reconsideration may also be necessary, since unstable or weakly formed preferences may evolve as the procedure advances.

This approach is actually a natural one. One would instinctively apply an informal version of it to address many daily decision problems, where one acts as both decision maker and analyst. This was the case of the father who asked himself a series of questions regarding the car that his family should purchase (see Chapter 1). In the beginning, the father only made preference judgments relative to automobile models with which he was very familiar. After this first stage of dialogue (with himself), he proceeded to a processing stage in which he tried to learn more about how certain models with which he was unfamiliar related to his preferences. He tried to discover, for example, which models performed better on certain aspects (e.g., fuel consumption, power, ...) than a model he knew well, while not performing significantly worse on other aspects (e.g., purchase price, seating capacity, ...). In this way he enriched the set of models that were familiar, and he was ready to begin a new dialogue with himself to elicit other preference judgments. This new dialogue would be based on new information and could very well force him to reconsider some of his earlier judgments or to pursue further his investigation of new automobile models. If he would not tire of the process or let himself be influenced by others, he would begin forming convictions that he did not have at the outset.

In this example, the processing and dialogue stages are not clearly separated. Moreover, their contents and their connections are left to the decision maker, who happens to be the analyst. Whether or not the analyst and decision maker are the same person, in more complex problems it becomes useful, even necessary, to describe formally (i.e., trying to anticipate all the possibilities) the contents of the two stages and how they are to interact and influence each other. In other words, it becomes necessary to build what we call an **interaction protocol.**[13]

11.4.2 Typical interaction protocols

The interaction protocol upon which this type of approach is based could be computerized, or it could be handled "personally." In either case, it must try to anticipate as exhaustively and rigorously as possible:

[13] *translator's note: At this point in the original French version, the Engine Assignment Example (Ex. 9) is continued to illustrate this idea. This presentation follows from a presentation of the example in Chapters 6, 8 and 9.*

a) how each stage of the dialogue will build upon the information gathered in previous stages; we refer to this part of the dialogue as the **explanation phase**;
b) the different possible types of questions, the types of information acquired in previous questions that will determine which subsequent questions will be asked, and how the questions will be formulated; we refer to this part of the dialogue stage as the **questioning phase**;
c) the different types of processing possible, how they are connected to the different types of questions, and how they will be influenced by the different types of answers provided; the results of these **processing phases** will be used to determine the characteristics of the following explanation phase.[14]

11.4.3 Important comments

In this third approach the actions in A may be descriptively enumerated, or they may be represented by points in an $\mathbb{R}^m$ space of m decision variables. The interaction protocol may be computerized or not. Whatever the setting, the analyst must pay particular attention to the following questions:

– How will the information that is to be communicated to the questionee be presented in the explanation phase – e.g., verbally, in tables, in diagrams, in graphs? How will it be presented so as to give the questionee the background to express the types of judgments anticipated without unnecessarily influencing these judgments?

– How will the questions be formulated in the questioning phase so that they are understood by the questionee and so that the judgments that they elicit produce useful information on her comprehensive preferences, or on how these preferences are forming, in the neighborhood of the small number of actions considered? Open-ended types of questions must be phrased so that one can extract with little ambiguity the local intercriterion information required to move on to the processing phase.

– What means or algorithms will be used to process the information collected so as to propose or direct the procedure towards what the questionee would consider a best compromise, an improved sorting, or a new basis for defining a preorder? To do so, that which is meaningful, essential, or new must be identified in the collected information and processed in such a way that it can be used to inform the questionee on new aspects or to formulate new propositions.

As mentioned in Section 11.4.1, the interaction protocol will include a process of trial and error and must thus be designed to allow it. Under these conditions, one might wonder what will indicate when the dialogue-processing sequence should be terminated.

[14] *translator's note: In the original French version, these points are developed further on Pages 378 and 379, and the end of the Product Composition example (Ex. 11) is presented as an illustration.*

Will it be when the parties involved discover a good compromise solution[15] in P.α, a good sorting in P.β, or a good ranking in P.γ? Or will it be when they become tired of the process? The answer will depend largely on whether a constructive or descriptive approach is adopted (see Roy, 1987).

In a constructive approach, the dialogue-processing sequence is seen as a tool to help progressively develop compromise solutions, sortings, or rankings. It is meant to take only temporary viewpoints that can be reconsidered at any moment. Therefore, the sequence of interactions can be stopped whenever the questionee feels either satisfied or frustrated. It can also be stopped by the questioner: The analyst may determine that enough relevant information has been obtained, or the computer program may reach a dead end if the questionee does not wish to reconsider previous judgments that lead the process to an impasse. It should be clear that one cannot think of the process "converging" in these cases. There would be no way of characterizing beforehand a target to which the procedure should lead the questionee or the questioner, since some of the "data" – primarily the comprehensive preferences, but also the constraints – are not fixed before the procedure begins; rather they are developed, modified, and refined as the solution is being constructed. That is, the data available at the beginning of the process are insufficient to define something towards which the procedure can converge.

In a descriptive approach, comprehensive preferences and the constraints on the problem are not supposed to be developed or transformed during the operation of the procedure. They are supposed to preexist in a stable state. The interactions are meant to reveal these preferences without altering them, even if this revelation occurs in a somewhat chaotic manner, often requiring several attempts to elicit the same information. The dialogue-processing sequence, then, can be considered as trying to discover a compromise solution or solutions (in the sense of optima), an ideal sorting, or a preorder that reflects the questionee's preferences exactly. In the descriptive approach, therefore, one should want the interaction protocol to converge on something that helps the decision process progress. This convergence could, of course, be hindered by what may seem like steps in the wrong direction because of difficulties involved with expressing or discovering preferences. It is only when this procedure reaches its goal or is sufficiently close to it, however, that the procedure should be ended.

The reader will find a description of the main interaction protocols in Vanderpooten and Vincke (1989), Steuer and Gardiner (1990), Roy and Bouyssou (1993, chapter 7).

To conclude, OA3 is based on two fundamental positions:

- giving primary importance to local judgments that deal with a very small number of actions, without considering any explicit rule that tries to aggregate, even partially or temporarily, the performance levels;

[15] *Because of the interactive and iterative nature of this third approach, the actions eventually selected in the case of problematic P.α usually represent compromises, or satisficing solutions as they were called in OA2. Unlike in OA1, they are rarely optima.*

- making explicit a protocol that organizes the interaction between the decision maker or other actors in the decision process and the analyst or a computer (questionee and questioner) so as to allow the recommendation to emerge for the problematic considered.

At times, it may be useful to weaken the first position in an attempt to combine this approach with OA1 or OA2. This, of course, would imply that the second positions in those approaches, presented in Sections 11.2.3 and 11.3.3, respectively, would likewise have to be weakened. PREFCALC (see Jacquet-Lagrèze and Shakun, 1984; Jacquet-Lagrèze, 1990), for example, represents one such combination in which OA1 and OA3 are combined under the hypothesis of an additive aggregation function V (see Section 11.2.2).

One could imagine other operational approaches than those presented in this chapter. The most common ones used in application, however, are based directly on OA1, OA2, and OA3, or as is the case with PREFCALC, on a simple combination of two of them.

Chapter 12

SPECIFIC DIFFICULTIES IN CHOICE, SORTING, AND RANKING PROBLEMATICS

SUMMARY

In the first section we discuss issues related to the choice of operational approach. Just as when choosing the problematic, the analyst may often hesitate among several possibilities. We propose that neither the choice of problematic nor the characteristics of A should systematically influence the choice of the operational approach. The influence will more likely come from the general environment surrounding the problem, the analyst's relation with the other actors, how the analyst will fit in the decision process, and the analyst's training.

We summarize general paths the analyst might take to overcome difficulties arising from: dependence among actions (e.g., exclusions, redundancies, complementarities, synergies) in Section 12.2; multiple scenarios in Section 12.3; conflicting value systems among actors in Section 12.4; hesitations on the roles that the various criteria play in directing a strategic decision in Section 12.5; and poorly defined actions with difficult to evaluate performance levels in Section 12.6. Although there could be other difficulties in practical applications, these common ones indicate the complexity of the problem. No decision aiding methodology can provide a recipe of steps. Its contribution will rather come from guiding the thought process by placing the various factors and relations in a systematic framework, thereby giving the decision maker and other actors improved insight into the problem at hand.

12.1 CHOOSING THE OPERATIONAL APPROACH

None of the operational approaches presented in Sections 11.2-11.4 can be considered best for all applications. Neither is it possible to develop a systematic correspondence between the operational approach to be employed and the type of application. Combining two of these approaches may be helpful in certain situations, but even combinations of approaches will not be able to address all eventualities.

Still, some approach must be chosen when faced with a specific application. The analyst has as much freedom in choosing the operational approach as he had in choosing the problematic for the application. One may find good reasons to reject some of the operational approaches for some of the 12 reference examples developed throughout this book. In many of the examples, however, a choice of one of the approaches would probably represent an individual preference that would be difficult to support when faced with arguments for a different approach.

In Table 12.1, we present the approaches that we felt most appropriate for each of the 12 reference examples. In light of what we said above, we felt that explaining why the indicated approach was chosen for the given example would lead to discussions too lengthy to include here. Table 12.1, like Table 6.2, leads to two important observations:

1) The choice of operational approach is not influenced by the problematic; each of the nine possible combinations (i.e., three problematics times three operational approaches) appears at least twice in Table 12.1.

2) The choice of operational approach is not significantly influenced by whether A has a comprehensive or fragmented conception or whether it is considered stable or not; considering Tables 12.1 and 6.2 together shows that all the combinations are present.

Most of the decisions made in everyday operations are based on somewhat simple reasoning with not entirely explicit criteria and without a formal decision aiding procedure. The processes often used in these decisions would probably be classified as belonging to the "local, interactive" approach OA3. In those cases where even a slightly formal procedure is introduced, however, OA3 is found much less often. Many specific procedures based on OA3 have been proposed, but with few exceptions they almost all consider P.α and deal with a set of actions that can be described by real-valued decision variables.[1] Therefore, analysts wishing to choose this approach will be forced to innovate, since they will find few documented interaction protocols that apply to the characteristics of the problems being considered and that have been validated under conditions similar to those at hand. The "synthesizing criterion" approach OA1 appears to be chosen most often. It is the oldest of the three approaches and is consequently the approach that is most widely taught and studied from a theoretical perspective. The "outranking" approach OA2 appeared only in 1966 with the introduction of ELECTRE I. Even though this approach has generated many other methods and been applied to numerous applications in several countries, relatively few of these have actually been published and, therefore, widely disseminated.

The choice of operational approach should be made by considering what the study hopes to accomplish while taking into account the progress of the decision process, the personalities of the client, decision maker, and relevant actors, and the type of access that the analyst will have to these individuals to explain his approach, gather information, and communicate the eventual results. More technical aspects, such as the number of actions and criteria considered, the degree to which the criteria conflict, and the degree of confidence in predicting the performance levels, can also play a role in this choice. By considering these and perhaps other factors in light of the comments made in Sections 11.2.3, 11.3.3, and 11.4.3, and by keeping in mind the options forming the basis of each of the three approaches, the analyst will be able to judge the advantages and disadvantages of each approach and either choose the one best suited to the problem considered or determine an attractive mixture of the approaches.

---

[1] *For exceptions see Vanderpooten and Vincke (1989), and untranslated continuation of Example 11: Fabrication d'un produit, Section 11.4.2, pp. 379-382 in original, French version of this book.*

Table 12.1: Synthesis of the characteristics of the 12 reference examples (see, also, Table 6.2)

Number, Title, Study Phase	Problematic	Operational Approach	Sections
Ex. 1. Site Selection: a) 1st phase b) 2nd phase	 β α	 3 (or 2) 2	3.1, 6.1.1, 11.3.2
Ex. 2. Commuter Rail Line: a) 1st phase b) 2nd phase	 β γ	 3 1 (or 2)	3.1, 5.2.2, 6.1.3, 8.2.3.1, , 8.2.4, 9.2.2.1
Ex. 3. Agricultural Development	α	1 (or 3)	3.2, 5.1.2, 6.1.1, 8.1.5, 8.2.3.2, 11.2.1
Ex. 4. Water Resource Planning: a) 1st phase b) 2nd phase	 β α	 2 (3 or 2)	3.2, 6.2.2, 8.2.3.1
Ex. 5. Media Planning	γ	2	3.3, 6.1.3, 8.1.5, 8.2.2.1, 10.1
Ex. 6. Research Projects	γ	1	3.4, 5.1.2, 6.1.3, 8.1.5, 8.2.2.2, 8.2.3.1
Ex. 7. Industrial Development	β (or γ)	2	3.4, 5.1.1, 6.1.2
Ex. 8. Airport Operations: a) 1st phase b) 2nd phase	 γ (or β) α	 1 1 (or 2)	3.5, 6.2.2, 8.2.4, 12.2
Ex. 9. Engine Assignment	β (or α or γ)	3 (or 1)	3.5, 5.2.2
Ex. 10. Application Packages	β	2	3.6, 6.1.2, 7.2.4, 9.4.1
Ex. 11. Product Composition: a) 1st case b) 2nd case	 α α	 1 (or 3) 3	3.7, 5.2.2, 6.1.1, 9.1.1
Ex. 12. Plant Organization	γ or mixed)	3	3.7, 6.2.2, 7.2.4, 9.4.2

12.2 PROBLEMS WITH NON-INDEPENDENT ACTIONS

Developing a plan based on the combination of fragmentary actions in a set A (Examples 5, 6, and 8) will very likely run into difficulties associated with incompatibilities, redundancies, complementarities, and synergistic effects in fragment combinations. If such effects did not exist, the plan could usually be developed in conjunction with some ranking procedure: One would develop the plan by choosing actions consecutively from the top of the complete or partial preorder until some predefined limit on the number of actions to be included or on some budget (financial, resource, time for implementation) was reached, or in the case of a ranking based on the unique synthesizing criterion, until the marginal utility or profitability fell below some value. When the actions in A are not independent, however, constructing a plan based on some defined preorder is not as simple.

Because of the dependencies among the fragmentary actions of A, the perspective is changed from reasoning on a set A to one of reasoning on a set A^* whose elements are combinations of elements of A that are sufficiently coherent with respect to the dependencies. The problem, then, becomes one of constructing A^* and evaluating its actions. This can be difficult, since the number of actions in A^* can be extremely large. One solution is to consider a preorder defined on A and the various dependencies among the fragmented actions (see, e.g., Danila, 1980, 1983; Jacquet-Lagrèze, 1995) to generate a set of plans A^* that, in principle, are both realistic and attractive.

Example 8: Airport Operations (from Section 8.2.4)

Recall from Section 6.2.2 that the first phase of the study was concerned with ranking roughly one hundred or so fragmented actions of A in accord with aspects a)-g) mentioned in Section 3.5. Assume that the analyst used ELECTRE III (see Roy, 1978; Vincke, 1992), PREFCALC (see Jacquet-Lagrèze, 1990), or some type of cost-benefit method (see Zoller and Beguin, 1992) to develop a complete order that could be used to begin the second phase, although the approach could be easily generalized to the case of a complete preorder or even a partial preorder. We could use a_1 to denote the first action in the order, a_2 the second action in the order, and so on, and we shall say that a_i is of rank i. The second phase of the study, then is concerned with:

– developing a set A^* of plans that are considered attractive;
– selecting one or a few best plans of A^*.

For various reasons, some of which were discussed in Section 8.2.4, evaluating a plan is more complicated than simply evaluating the various fragments of A that constitute the plan. Therefore, the number of combinations of fragments that become elements of A^* must be kept relatively small without throwing away combinations that would represent particularly attractive plans.[2]

(end of Example 8)

[2] *translator's note: The details of one way of accomplishing this are presented in the continuation and end of this example on Pages 391-392 in the original French version.*

12.3 PROBLEMS WITH MULTIPLE SCENARIOS

Consider a decision aiding situation that must take into account a set S of scenarios. This means that the performance level $g_j(a)$ for all or any combination of criteria g_j in the coherent criteria family F considered can vary with a scenario $s \in S$. Therefore, we denote the performance level $g_j(a, s)$, and the implication is that there is a performance tableau for each scenario. Since one cannot predict which scenario will occur in the future, the client expects the eventual recommendation to take the various scenarios into account, but he or she also expects one recommendation and not one for each scenario. To provide such a recommendation, the analyst may run into two types of difficulties:

- how to account for the fact that the different scenarios are not necessarily **equally credible** and, as a consequence, the relative importance to give to the particularly disastrous or beneficial effects of certain scenarios;
- how to account for the fact that all of the actions are not **equally flexible** and, as a consequence, the relative importance to give to the adaptability of options to the possible realization of various scenarios.

To address the unequal flexibility of the actions, one might sometimes be able to distinguish a relatively irreversible part of each action and another part that could be completely reconsidered or slightly modified, depending on future events. Certain actions might, therefore, stand out as being more flexible than others with respect to the uncertainties characterizing the different scenarios.

A multicriterion methodology presents several ways of handling the information contained in the performance tableaus that correspond to the different scenarios. They boil down, however, to essentially two not necessarily exclusive ways of attacking the problem.

a) Eliminating the performance level dispersion caused by the scenarios

For each criterion g_j and each action a, the analyst reduces the set of performance levels $\{g_j(a, s) / s \in S\}$ to a unique number $g_j(a)$.[3] For example, he could use a point reduction method (Section 9.2.2.1). This in turn reduces the set of performance tableaus to one, which corresponds to the same type of problem as the initial one, but without scenarios.

b) Synthesizing the results corresponding to each scenario

The analyst determines the choice, sorting, or ranking that would be recommended in each of the scenarios of S. He then compares the results and tries to synthesize them. In the case of problematic P.γ, if the result obtained for each scenario s is a complete preorder, one could consider the preorders to compare the actions in A according to the same significance axis but in the context of a particular scenario. This would be exactly

---

[3] *Reducing the set to a small set of numbers that correspond to a splitting of the significance axis of criterion g_j would also be possible.*

as if a different criterion were associated with each scenario, and the synthesis would be obtained by treating this newly formed problem as one conforming to problematic P.γ. The analyst could proceed in a similar fashion in the case of problematics P.α and P.β – ranking the actions according to each scenario and then considering each ranking as a criterion; these criteria would constitute a new family in the problematic to be considered.[4]

12.4 PROBLEMS WITH CONFLICTING VALUE SYSTEMS

Many decision situations involve multiple actors, each with a distinct value system. In these cases, the value systems are rarely the same and can even conflict with respect to important elements during the process. The conflict may be due to any of a number of factors – e.g., different ethical or ideological beliefs, different specific objectives, or different roles within an organization. Whatever the origin of the conflicting value systems, they will usually affect the evolution of the decision process in ways that were not expected at the outset. Who can exert what influence, who has the power to take advantage of some opportunity, who has access to what information, and who is the most convincing actor will have an important impact on the final solution. With such factors at work, it is often difficult to determine the impact of a decision aiding study, even after the fact. Especially in these cases, one must be careful not to try to dictate a decision. The study may be requested by one of the actors to help rationalize her conduct or perhaps by a mediator trying to guide the decision process. In either case, the study should not pretend to be more than an aid to clarifying the various issues and impacts. If the study can contribute to reducing the confusion surrounding the process, its impact will be important.

The most important task in these types of situations is determining a coherent criterion family F that can then lead to the construction of a performance tableau that does not depend on the differences in the value systems. This would correspond to problematic P.δ. In the other problematics, the impacts of the criteria on the comprehensive preferences will be important. These do depend on the specific value systems and, therefore, the specific actors. One way to address this problem consists of considering the problem from each actor's (or each value system's) viewpoint, i.e., to consider that there are as many problems as there are actors. The comprehensive preferences of each actor are all handled with the same operational approach, but the intercriterion information is different for each. The different results are then used in the context of the same problematic – preferably P.α or P.β, since the ranking required of P.γ necessarily depends on the value system considered. From a technical or mechanical point of view, this approach is the same as that of Section 12.3b). The difference arises in the interpretation of the divergent results – as being generated either by different scenarios or by different value systems.

---

[4] *translator's note: The continuation and end of Example 2 in the original French version illustrates the application of these two ways of attacking the problem in the case of P.γ on Pages 394-397. The extension to P.α or P.β would be straightforward.*

When the study is conducted for the benefit of one of the actors, using some model to portray what underlies the position of the other actors can provide a clearer understanding of certain aspects of the problem and a better basis for arguing the actor's position. This is so even when the model possesses some arguable hypotheses. When the study is conducted for the benefit of a mediator, this type of comparative study can indicate where the conflicts occur. To which category each actor assigns each action of A or the rank each actor gives to each action might allow the mediator to eliminate certain actions from A and to identify ways to propose new actions and develop compromises. Finally, we note that any study may reveal something very different than what a given actor would like, and during the decision process the actor may act according to some tactical strategy that she would generally not follow if there were no conflict among the value systems.[5]

12.5 PROBLEMS WITH STRATEGIC HESITATIONS

In many strategic problems – problems dealing with financial (see Zopounidis, 1987; Colson and De Bruyn, 1989; Diakoulaki *et al.*, 1992), commercial, production and distribution strategies (see Lock, 1982; Siskos, 1986; Roberts and Urban, 1988), or strategies in which environmental aspects play a decisive role (see Maystre *et al.*, 1994; Paruccini 1994), for example – the analyst will find it very difficult to build a model of comprehensive preferences. In these problems even well-defined decision makers who accept the suitability of the criteria developed for the given problem will not have a clear understanding of their roles in developing comprehensive preferences from the criteria. They may, for example, be reluctant to think of compensating a poor performance level on one criterion by good performance levels on others or to think of the possibility of allowing for veto thresholds. This type of reluctance would exist in other types of problems as well, but it seems to take on greater importance in strategic problems where there are more risks from unforeseen events and where what is truly possible is harder to distinguish from what is not. The decision maker will generally expect the decision aid to indicate the "most justified" way of using the criteria to develop a strategy. More than in other problems, it will be ideological, political, cultural, or circumstantial factors that will affect the soundness of the role each criterion plays in aggregating the performance levels. Although the methodology developed in this book cannot address this problem directly, it can be useful in analyzing for what reasons and to what degree the strategies depend on the different forms of reluctance or hesitations that occur.

When trying to aid in this way, the analyst should first try to identify the points where the major forms of hesitation can be found. He then must develop the operational approach in such a way that it can account for these hesitations in the structural variants (i.e., the various functional forms of V or tests T) and in the parameters associated with

[5] *translator's note: The continuation and end of Example 4 is presented on Pages 398-401 at this point in the original, French version.*

the intercriterion information. The number of ways that the different structural variants and parameter values can be combined is very large, however, and the possibilities must be pruned in some fashion. Let M denote the set of indices corresponding to the various combinations of structural variants and parameter values to be considered after pruning the combination to a reasonable level. Using the same procedure (related to the problematic considered) with each of the combinations m leads to a set of results (e.g., a preorder on A) that we shall denote P_m. The analyst must then use the set $\{P_m \mid m \in M\}$ to bring about:

- a recommendation that can be supported by results corresponding to a sub-family M′ of M that is as large as possible;
- conclusions that will communicate how other combinations of structural variants and parameter levels not corresponding to those of M′ would modify this recommendation.

Only recently has research begun to develop systematic tools that would quickly assist in such an analysis of robustness resulting in what we like to call "quasi-invariants," i.e., configurations whose very nature may be poorly defined at the outset but that appear in the same or very similar form in most or a "sufficient majority" of the P_m's. For example, when considering the P_m's to be preorders, one could think of the quasi-invariants as being related to the relative place of two actions or groups of actions in the order (i.e., "a is almost always before a′") or of the relative ranking of an action or subset of actions (e.g., "at least one of a, a′, or a″, is almost always at the top of the order, and the others are either in the same class, immediately behind, or incomparable to it or them.") Roy and Hugonnard (1982) demonstrate the use of such quasi-invariants in a practical study, and Roy *et al.* (1986) outline a technique for such robustness analysis. In general, such an analysis will not only account for the strategic hesitations, but also for other factors of imprecision, uncertainty, and inaccurate determination of the consequences (more details are given in Roy and Bouyssou, 1993, p. 313-318).[6]

12.6 PROBLEMS WITH POORLY DEFINED SETS OF ACTIONS AND HARD-TO-ESTIMATE PERFORMANCE LEVELS

In some cases, decision aiding studies are requested because the actions of interest are so complex that it is difficult to formulate them and determine a set of actions to be considered. This complexity makes it especially difficult to determine a coherent criterion family and to value the actions according to the criteria of this family. Even in these situations a formal methodology, such as the one developed in this book, can be a useful guide.[7]

---

[6] *translator's note: The continuation and end of Example 7 is presented on Pages 403-404 at this point in the original French version as an illustration.*

[7] *translator's note: The continuation and end of Example 12 is presented on Pages 405-406 at this point in the original French version as an illustration.*

BIBLIOGRAPHY

[1] Abgueguen, R., *La sélection des supports de presse*, Robert Laffont, Paris, (1971).

[2] Ackoff, R.L., Optimization + objectivity = opt out, *European Journal of Operational Research*, (1977), 1, 1-8.

[3] Adams, E.W., Elements of a theory of inexact measurement, *Philosophy of Science*, (1965), 32, 205-228.

[4] Alexandre A., Barde, J.P., L'implantation des aéroports – Peut-on en évaluer les coûts et les avantages, *Analyse et Prévision*, (1973), XV, 385-420.

[5] Arrow, K.J., *Social choice and individual values*, 2nd edition, John Wiley and Sons, New York, (1963).

[6] Bana e Costa, C.A. (ed.), *Readings in multiple criteria decision aid*, Springer-Verlag, Berlin, (1990).

[7] Bana e Costa, C.A., Dinis das Neves, C., Describing and formalizing the evaluation process of Portuguese navy officers, in A.G. Lockett and G. Islei (eds.), *Improving Decision Making in Organisations*, LNEMS 335, Springer-Verlag, Berlin, (1989), 355-369.

[8] Bana e Costa, C.A., Rodrigues, J.C., Presentation of the school case-study: Evaluation of personnel – How to rank officers for promotion?, in C.A. Bana e Costa (ed.), *Readings in Multiple Criteria Decision Aid*, Springer-Verlag, Berlin, (1990), 639-647.

[9] Banville, C., Landry, M., Martel, J.M., Boulaire, C., A stakeholder's approach to MCDA, Université Laval, CRAEDO, *Working Paper* 93-79, (1993).

[10] Bell, D.E., Raiffa, H., Tversky, A. (eds.), *Decision making: Descriptive, normative and prescriptive interactions*, Cambridge University Press, (1988).

[11] Belton, V., Vickers, S., Use of a simple multi-attribute value function incorporating visual interactive sensitivity analysis for multiple criteria decision making, in C.A. Bana e Costa (ed.), *Readings in Multiple Criteria Decision Aid*, Springer-Verlag, Berlin, (1990), 319-334.

[12] Benzecri J.P. (ed.), *L'analyse des données – Tome 1 : La taxinomie, Tome 2 : L'analyse des correspondances*, Dunod, Paris, 4th edition (1984).

[13] Berge, C., *Graphs and Hypergraphs*, North Holland Publishing Company, Amsterdam, (1973).

[14] Berry, M., *Une technologie invisible ? L'impact des instruments de gestion sur l'évolution des systèmes humains*, Centre de Recherche en Gestion, Ecole Polytechnique, Paris, (1983).

[15] Bertier, P., Bouroche, J.M., *L'analyse des données multidimensionnelles*, PUF, Collection Systèmes-Décisions, Paris, (1975).

[16] Bertier, P., de Montgolfier, J., Comment choisir en tenant compte de points de vue non commensurables, *Analyse et Prévision*, (1971), XI, 521-548.

[17] Bétolaud, Y., Février, R., Conservation des forêts suburbaines et passage des autoroutes – L'exemple du tracé de l'autoroute A86-A87, *Revue Forestière Française*, (1973), XXV, n° 3, 179-200.

[18] Boudon, R., *Effets pervers et ordre social*, PUF, Paris, (1977).

[19] Bouyssou, D., Decision-aid and expected utility theory: A critical survey, in O. Hagen and F. Wenstøp (eds.), *Progress in Utility and Risk Theory*, D. Reidel Publishing Company, Dordrecht, (1984), 181-216.

[20] Bouyssou, D., On some properties of outranking relations based on a concordance-discordance principle, in A. Goicoechea, L. Duckstein, S. Zionts (eds.), *Multiple Criteria Decision Making*, Springer-Verlag, (1992), 93-106.

[21] Bouyssou, D., Vansnick, J.C., Noncompensatory and generalized noncompensatory preference structures, *Theory and Decision*, (1986), 21, 251-266.

[22] Brans, J.P., Mareschal, B., The PROMETHEE methods for MCAM: The PROMCALC, GAIA and BANKADVISER software, in C.A. Bana e Costa (ed.), *Readings in Multiple Criteria Decision Aid*, Springer-Verlag, Berlin, (1990), 216-252.

[23] Brans, J.P., Mareschal, B., Vincke, Ph., PROMETHEE: A new family of outranking methods in multicriteria analysis, in J.P. Brans (ed.), *Operational Research '84*, Elsevier Science Publishers B.V. (North-Holland), Amsterdam, (1984), 408-421.

[24] Chankong, V., Haimes, Y.Y., *Multiobjective decision making: Theory and methodology*, North-Holland Publishing Company, New York, (1983).

[25] Chapman, W.L., Bahill, A.T., Wymore, A.W., *Engineering modeling and design*, CRC Press, Boca Raton, FL, (1992).

[26] Charpentier, A.R., Jacquet-Lagrèze, E., La promotion de l'électricité dans l'industrie et l'utilisation de méthodes multicritères, in H. Thiriez and S. Zionts (eds.), *Multiple Criteria Decision Making*, LNEMS 130, Springer-Verlag, Berlin, (1976), 364-377.

[27] Christofides, N., *Graph theory – An algorithmic approach*, Academic Press Inc., London, (1975).

[28] Cleland, D.J., King, W.R., *System analysis and project management*, 3rd edition, McGraw Hill, (1983).

[29] Colson, G., De Bruyn, C., Integrated multiobjective portfolio management system, *Mathematical and Computer Modelling*, (1989), 12, 1359-1381.

[30] Condorcet, M.J.A.N.C. (Marquis de), *Essais sur l'application de l'analyse à la probabilité des décisions rendues à la pluralité des voix*, Paris, (1785).

[31] Crozier, M., Friedberg, E., *L'acteur et le système*, Editions du Seuil, Paris, (1977).

[32] Danila, N., *Méthodologie d'aide à la décision dans le cas d'investissements forts dépendants*, Université Paris-Dauphine, Thèse de 3e Cycle, (1980).

[33] Danila N., *Stratégies technologiques – Méthode d'évaluation et de sélection de projets de recherche*, Fondation Nationale pour l'Enseignement de la Gestion des Entreprises, Paris, (1983).

[34] D'Avignon, G.R., Sauvageau, M., L'aide multicritère à la décision : Un cas d'intégration de critères techniques, économiques et environnementaux à Hydro-Québec, *RAIRO Recherche Opérationnelle*, to appear, (1996).

[35] Déry, R., Landry, M., Banville, C., Revisiting the issue of model validation in OR: An epistemological view, *European Journal of Operational Research*, (1993), 66, 168-183.

[36] Despontin, M., Dynamic optimization in a regional econometric model for Belgium, in J.P. Brans (ed.), *Operational Research '81*, North-Holland Publishing Company, (1981), 283-294.

[37] Despontin, M., Regional multiple objective quantitative economic policy: A Belgian model, *European Journal of Operational Research*, (1982), 10, 82-89.

[38] Diakoulaki, D., Mavrotas, G., Papayanakis, L., A multicriteria approach for evaluating the performance of industrial firms, *OMEGA*, (1992), 20, 467-474.

[39] Dubois, D., Prade, H., *Fuzzy sets and systems: Theory and applications*, Academic Press, New York, (1980).

[40] Edwards, W., (ed.), *Utility theories: Measurements and applications*, Kluwer Academic Publishers, (1992).

[41] Emsellem, B., La nécessaire limitation du rôle de l'informatique dans l'aide à la décision, in Actes du Colloque *Informatique, Automatique et Sciences des Organisations*, Rocquencourt, France, 14-16 janvier 1976, Editions IRIA, Paris, (1976), 93-106.

[42] Fayette, J., Programmation linéaire à plusieurs fonctions d'objectif : Application à la planification agricole en République de Corée, Université Jean Moulin, Institut d'Administration et de Gestion des Entreprises, *Document GRAE*, Lyon, France, (1975).

[43] Fishburn, P.C., *Utility theory for decision-making*, John Wiley and Sons, New York, (1970).

[44] Fishburn, P.C,. A survey of multiattribute/multicriteria evaluation theories, in S. Zionts (eds.), *Multiple Criteria Problem Solving*, LNEMS 155, Springer-Verlag, (1978), 181-224.

[45] French, S., *Decision theory: An introduction to the mathematics of rationality*, Ellis Horwook, Chichester, (1986).

[46] Giordano, J.L., Suquet, J.C., On multicriteria decision making: An application to a workshop organization problem, in Proceedings of *EURO II, Second European*

Congress on Operations Research, Stockholm, Sweden, November 29-December 1, 1976, North-Holland Publishing Company, (1976), 181-192.

[47] Goicoechea, A., Duckstein, L., Zionts S., (eds.), *Multiple criteria decision making*, Springer-Verlag, (1982).

[48] Goicoechea, A., Hansen, D.R., Duckstein, L., *Multiobjective decision analysis with engineering and business applications*, John Wiley and Sons, Chichester, (1982).

[49] Gregory, R., Lichtenstein, S., Slovic, P., Valuing environmental resources: A constructive approach, *Journal of Risk and Uncertainty*, (1993), 7, 177-197.

[50] GRETU, *Une étude économique a montré... Mythes et réalités des études de transport*, Cujas, Paris, (1980).

[51] Hatchuel, A., Molet, H., Rational modeling in understanding and aiding human decision making: About two case studies, *European Journal of Operational Research*, (1986), 24, 180-188.

[52] Hatchuel, A., Weil, B., Experts in organizations – A knowledge-based perspective on organizational change, de Gruyter, (1995).

[53] Hershey, J.C., Kunreuther, H.C., Schoemaker, P.J.H., Sources of bias in asessment procedures for utility functions, *Management Science*, (1982), 28, 936-954.

[54] Heurgon, E., Relationships between decision making process and study process in OR intervention, *European Journal of Operational Research*, (1982), 10, 230-236.

[55] Hinloopen, E., Nijkamp, P., Rietveld, P., Qualitative discrete multiple criteria choice models in regional planning, *Regional Science and Urban Economics*, (1983), 13, 77-102.

[56] Hirsch, G., Jacquet-Lagrèze, E., Moscarola, J., Définition et illustration de quelques concepts utiles à la description et la maîtrise d'un processus de décision, in Actes du Congrès AFCET *Modélisation et Maîtrise des Systèmes Techniques, Economiques, Sociaux*, Editions Hommes et Techniques, Paris, (1977).

[57] Holl, J.C., Plas, J.P., Riou, P., *Le choix d'investissement dans l'entreprise*, PUF, Paris, (1973).

[58] Jacquet-Lagrèze, E., *Systèmes de décision et acteurs multiples – Contribution à une théorie de l'action pour les sciences des organisations*, Université Paris-Dauphine, Thèse d'Etat, (1981).

[59] Jacquet-Lagrèze, E., Interactive assessment of preferences using holistic judgements: The PREFCALC system, in C.A. Bana e Costa (ed.), *Readings in Multiple Criteria Decision Aid*, Springer-Verlag, Berlin, (1990), 336-350.

[60] Jacquet-Lagrèze, E., An application of the UTA discriminant model for the evaluation of R & D projects, in P.M. Pardalos, Y. Siskos, C. Zopounidis (eds.), *Advances in Multicriteria Analysis*, Kluwer Academic Publishers, (1995), 203-211.

[61] Jacquet-Lagrèze, E., Marchet, J.C., Description d'un processus de décision – Extension d'une station d'épuration d'eaux usées, Université Paris-Dauphine, *Cahier du LAMSADE* 21, (1978).

[62] Jacquet-Lagrèze, E., Roy, B., Moscarola, J., Hirsch, G., Description d'un processus de décision – I : Quelques concepts, Université Paris-Dauphine, *Cahier du LAMSADE* 13, (1978).

[63] Jacquet-Lagrèze, E., Shakun, M., Decision support systems for semistructured buying decisions, *European Journal of Opeational Resarch*, (1984), 16, 48-56.

[64] Jaffray, J.Y., Cohen, M., An experimental study of multiplicative separable representation of preference under risk, Communication presented at *EURO VII, 7th European Congress on Operations Research*, Bologna, Italy, (1985).

[65] Keeney, R.L., *Value focused thinking: A path to creative decision making*, Harvard University Press, (1992).

[66] Keeney, R.L., Raiffa H., *Decisions with multiple objectives: Preferences and value tradeoffs*, John Wiley and Sons, New York, (1976).

[67] Khouadja, H., Roy, B., Comment choisir la localisation d'une grande école d'ingénieurs : l'exemple de l'Ecole Supérieure d'Ingénieurs de Marseille, Université Paris-Dauphine, *Cahier du LAMSADE*, (1975), 1.

[68] Klir, G.J., *Trends in general system theory*, John Wiley and Sons, New York, (1972).

[69] Landry, M., Oral, M., In search of a valid view of model validation for operations research, *European Journal of Operational Research*, (1993), 66, 161-167.

[70] Larichev, O.I. Cognitive validity in design of decision-aiding techniques, *Journal of Multi-Criteria Decision Analysis* (1992), 1, 127-138.

[71] Larichev, O.I., Moshkovich, H.M., An approach to ordinal classification problems, *International Transactions in Operational Research*, (1994), 1, 375-385.

[72] Le Boulanger, H., Roy, B., L'entreprise face à la sélection et à l'orientation des projets de recherche : La méthodologie en usage dans le Groupe SEMA, *Revue METRA*, (1970), VII, 641-669.

[73] Le Moigne, J.L., *La théorie du système général – Théorie de la modélisation*, PUF, Paris, (1977).

[74] Lesourne, J., *A theory of the individual for economic analysis*, Volume 1, North-Holland Publishing Company, (1977).

[75] Lévy-Lambert, H., Guillaume, H., *La rationalisation des choix budgétaires – Technique d'analyse*, PUF, Paris, (1971).

[76] Lock, A.R., A strategic business decision with multiple criteria: The Bally men's shoe problem, *Journal of the Operational Research Society*, (1982), 33, 327-332.

[77] Luce, R.D., Semi-orders and a theory of utility discrimination, *Econometrica*, (1956), 24, 178-191.

[78] Lyden, F.J., Miller, E.G., *Planning-programming-budgeting – A systems approach to management*, Markham Publishing Co., Chicago, (1967).

[79] Major, W., Moscarola, J., Analyse d'un processus de décision - Quelques conclusions à partir d'une étude de cas sur la sélection de projets d'expérience en physique des hautes énergies, Université Paris-Dauphine, *Cahier du LAMSADE*, (1979), 22.

[80] Martel, J.M., D'Avignon, G.R., Couillard, J., A fuzzy outranking relation in multiple criteria decision making, *European Journal of Operational Research*, (1986), 25, 258-271.

[81] Massaglia, R., Ostanello, A., N-TOMIC: A decision support for multicriteria segmentation problems, in P. Korhonen, A. Lewandowski, J. Wallenius (eds.), *Multicriteria Decision Support*, LNEMS 356, Springer-Verlag, Berlin, (1991), 167-174.

[82] Matarazzo, B., Multicriterion analysis of preferences by means of pairwise actions and criterion comparisons (MAPPAC), *Applied Mathematics and Computation*, (1986), 18, 119-141.

[83] Matarazzo, B., Preference ranking global frequencies in multicriterion analysis (PRAGMA), *European Journal of Operational Research*, (1988), 36, 36-50.

[84] Maystre, L.Y., Pictet, J., Simos, J., *Méthodes multicritères ELECTRE – Description, conseils pratiques et cas d'application à la gestion environnementale*, Presses Polytechniques et Universitaires Romandes, Lausanne, (1994).

[85] McCord, M.R., *Empirical demonstration of utility dependence on the fundamental assessment parameters: Revised assessment methodology*, MIT, Ph.D. Dissertation, (1983).

[86] McCord, M.R., Franzese O., Sun, X.D., Attribute importance in the supply of aeromedical service, *Transportation Research Record* 1359, (1992), 95-102.

[87] McCord, M.R., Franzese O., Sun X.D., Multicriteria analysis of aeromedical fleet expansion, *Journal of Applied Mathematics and Computing*, (1993), 54, 2 & 3, 101-129.

[88] McCord, M.R., de Neufville, R., Empirical demonstration that expected utility decision is not operational, in B.P. Stigum and F. Wenstøp (eds.), *Foundations of Utility and Risk Theory with Application*, D. Reidel Publishing Company, (1984), 181-199.

[89] Miser, H.J., A foundational concept of science appropriate for validation in operational research, *European Journal of Operational Research*, (1993), 66, 204-215.

[90] de Montgolfier, J., Bertier, P., *Approche multicritère des problèmes de décision*, Editions Hommes et Techniques, Paris, (1978).

[91] Morin, E., *La méthode – I : La nature de la nature*, Editions du Seuil, Paris, (1977).

[92] Moscarola, J., Roy, B., Procédure automatique d'examen de dossiers fondée sur une segmentation trichotomique en présence de critères multiples, *RAIRO Recherche Opérationnelle*, (1977), 2, 145-173.

[93] Mousseau, V., Elicitating information concerning the relative importance of criteria, in P.M. Pardalos, Y. Siskos, C. Zopounidis (eds.), *Advances in Multicriteria Analysis*, Kluwer Academic Publishers, (1995), 17-43.

[94] de Neufville, R., *Applied systems analysis*, McGraw-Hill, New York, (1991).

[95] Norese, M.F., Ostanello A., Identification and development of alternatives: Introduction to the recognition of process typologies. In: Lockett A.G. and Islei G., eds. *Improving Decision Making in Organizations*, LNEMS 335, Springer-Verlag, Berlin, (1989), 112-123.

[96] Optner, S.L., *L'analyse des systèmes et les problèmes de gestion*, Dunod, Paris, Collection SIGMA, (1968).

[97] Ostanello, A., Action evaluation and action structuring: Different decision aid situations reviewed through two actual cases, in C.A. Bana e Costa (ed.), *Readings in Multiple Criteria Decision Aid*, Springer-Verlag, Berlin, (1990), 36-57.

[98] Oury, J.M., *Economie politique de la vigilance*, Calmann-Lévy, Paris, (1983).

[99] Ozernoy, V.M., Multiple criteria decision making in the USSR: A survey, *Naval Research Logistics*, (1988), 6, 543-566.

[100] Paelinck, J., QUALIFLEX: A flexible multiple criteria method, *Economic Letters*, (1978), 3, 193-197.

[101] Paruccini, M., (ed.), *Applying multiple criteria aid for decision to environmental management*, Kluwer Academic Publishers, (1994).

[102] Perny, P., Roy, B., The use of fuzzy outranking relations in preference modelling, *Fuzzy Sets and Systems*, (1992), 49, 33-53.

[103] Piaget, J. (ed.), *Logique et connaissance scientifique*, Encyclopédie de la Pléiade, NRF, Paris, (1973).

[104] Pirlot, M., A report on the study of the Portuguese navy case, in C.A. Bana e Costa (ed.), *Readings in Multiple Criteria Decision Aid*, Springer-Verlag, Berlin, (1990), 648-660.

[105] Pomerol, J.Ch., Barba-Romero, S., *Choix multicritère dans l'entreprise – Principes et pratique*, Hermès, Paris, (1993).

[106] Ponssard, J.P., *Logique de la négociation et théorie des jeux*, Les Editions d'Organisation, Paris, (1977).

[107] Popper, K.R., *Objective knowledge: An evolutionary approach*, 5th impression (revisited), Clarendon Press, Oxford, (1979).

[108] Raiffa, H., *Decision analysis – Introductory lectures on choice under uncertainty*, Addison-Wesley, Reading, (1970).

[109] Régnier, A., *Les infortunes de la raison*, Editions du Seuil, Paris, (1966).

[110] Riveline, C., Nouvelles approches des processus de décision – Les apports de la recherche en gestion, *Futuribles*, (1983), 72, 65-77.

[111] Rizzi, M., Plausibility relations and multiple expected utilities in decision making under uncertainty, *European Journal of Operational Research*, (1984), 16, 358-364.

[112] Robert, P., *Dictionnaire alphabétique et analogique de la langue française*, Société du Nouveau Littré, Paris, (1968).

[113] Roberts, J.H., Urban, G.L., Modeling multiattribute utility, risk and belief dynamics for new consumer durable brand choice, *Management Science*, (1988), 34, 167-185.

[114] Roubens, M., Preference relations on actions and criteria in multicriteria decision making, *European Journal of Operational Research*, (1982), 10, 51-55.

[115] Roubens, M., Vincke, Ph., *Preference modelling*, LNEMS 250, Springer-Verlag, Berlin, (1985).

[116] Roy, B., Classement et choix en présence de points de vue multiples (La méthode ELECTRE), *RIRO*, (1968) 8, 57-75.

[117] Roy, B., *Algèbre moderne et théorie des graphes orientées vers les sciences économiques et sociales*, Dunod, Paris, 2 volumes, (1969-1970).

[118] Roy, B., Problems and methods with multiple objective functions, *Mathematical Programming*, (1971), 1, 239-266.

[119] Roy, B., How outranking relations help multiple criteria decision making, in J.L. Cochrane and M. Zeleny (eds.), *Multiple Criteria Decision Making*, University of South-Carolina Press, (1973), 179-201.

[120] Roy, B., Critères multiples et modélisation des préférences (L'apport des relations de surclassement), *Revue d'Economie Politique*, (1974), 84, 1-44.

[121] Roy, B., From optimization to multicriteria decision aid: Three main operational attitudes, in H. Thiriez and S. Zionts (eds.), *Multiple Criteria Decision Making*, Springer-Verlag, LNEMS 130, (1976), 1-34.

[122] Roy, B., Partial preference analysis and decision-aid: The fuzzy outranking relation concept, in David E. Bell, Ralph L. Keeney and Howard Raiffa (eds.), *Conflicting Objectives in Decisions*, John Wiley and Sons, (1977), 40-75.

[123] Roy, B., ELECTRE III : Un algorithme de classements fondé sur une représentation floue en présence de critères multiples, *Cahier du CERO*, (1978), 20, 3-24.

[124] Roy, B., A multicriteria analysis for trichotomic segmentation problems, in P. Nijkamp and J. Spronk (eds.), *Multiple Criteria Analysis – Operational Methods*, Gower Press, (1981), 245-257.

[125] Roy, B., *Méthodologie multicritère d'aide à la décision*, Economica, Paris, (1985).

[126] Roy, B., Meaning and validity of interactive procedures as tools for decision making, *European Journal of Operational Research*, (1987), 31, 297-303.

[127] Roy, B., Main sources of inaccurate determination, uncertainty and imprecision in decision models, in B. Munier and M.F. Shakun (eds.), *Compromise, Negotiation and Group Decision*, Theory and Decision Library, D. Reidel Publishing Company, Dordrecht, (1988), 43-62.

[128] Roy, B., Decision-aid and decision-making, *European Journal of Operational Research*, (1990), 45, 324-331.

[129] Roy, B., The outranking approach and the foundations of ELECTRE methods, *Theory and Decision*, (1991), 31, 49-73.

[130] Roy, B., Decision science or decision-aid science?, *European Journal of Operational Research*, (1993), 66, 184-203.

[131] Roy, B., On operational research and decision-aid, *European Journal of Operational Research* (1994), 73, 23-26.

[132] Roy, B., Bertier, P., La méthode ELECTRE II, une application au média-planing, in M. Ross (ed.), *OR '72*, North-Holland Publishing Company, Amsterdam, (1973), 291-302.

[133] Roy, B., Bouyssou, D., Comparison of two decision-aid models applied to a nuclear plant siting example, *European Journal of Operational Research*, (1986), 25, 200-215.

[134] Roy, B., Bouyssou, D., *Aide multicritère à la décision : méthodes et cas*, Economica, Collection Gestion, Paris, (1993).

[135] Roy, B., Hugonnard, J.Ch., Ranking of suburban line extension projects on the Paris metro system by a multicriteria method, *Transportation Research*, (1982), 16A, 301-312.

[136] Roy, B., Mousseau, V., A theoretical framework for analysing the notion of relative importance of criteria, *Journal of Multi-Criteria Decision Analysis*, (1996).

[137] Roy, B., Présent, M., Silhol, D., A programming method for determining which Paris metro stations should be renovated, *European Journal of Operational Research*, (1986), 24, 318-334.

[138] Roy, B., Slowinski, R., Criterion of distance between technical programming and socio-economic priority, *RAIRO Recherche Opérationnelle*, (1993), 27, 45-60.

[139] Roy, B., Slowinski, R., Treichel, W., Multicriteria programming of water supply systems for rural areas, in K.W. Hipel (ed.), *Multiple Objective Decision Making*

in *Water Resources*, AWRA (American Water Resources Association) Monograph Series, (1992), 18, 13-32.

[140] Roy, B., Vanderpooten, D., The European School of MCDA: Emergence, basic features and current works, *Journal of Multi-Criteria Decision Analysis*, (1996).

[141] Roy, B., Vincke, Ph., Pseudo-orders: Definition, properties and numerical representation, *Mathematical Social Sciences*, (1987), 14, 263-274.

[142] Schärlig A. *Décider sur plusieurs critères – Panorama de l'aide à la décision multicritère*, Presses Polytechniques Romandes, Collection Diriger l'Entreprise, Lausanne (1985).

[143] Schwartz, T., Rationality and the myth of the maximum, *Noûs*, (1972), 6, 97-117.

[144] Seo, F., Sakawa, M., *Multiple criteria decision analyusis in regional planning: Concepts, methods and applications*, D. Reidel Publishing Company, (1988).

[145] Sfez, L., (1973) : *Critique de la décision*, Presses de la Fondation Nationale des Sciences Politiques, Paris, (1973).

[146] Simon, H.A., *The sciences of the artificial*, 2nd Edition, The MIT Press, Cambridge, Mass., (1981).

[147] Siskos, J., Evaluating a system of furniture retail outlets using an interactive ordinal regression method, *European Journal of Operational Research*, (1986), 23, 179-193.

[148] Slovic, P., From Shakespeare to Simon: Speculations – and some evidence – about man's ability to process information, Oregon Research Institute, *Research Monograph*, (1972), 12.

[149] Spetzler, C.S., Von Holstein, C.A.S., (1975) : Probability encoding in decision analysis, *Management Science*, (1975), 22, 340-358.

[150] Steuer, R.E., *Multiple criteria optimization: Theory, computation, and application*, John Wiley and Sons, New York, (1986).

[151] Steuer, R.E., Gardiner, L.R., Interactive multiple objective programming: Concepts, current status and future directions, in C.A. Bana e Costa (ed.), *Readings in Multiple Criteria Decision Aid*, Springer-Verlag, Berlin, (1990), 413-444.

[152] Suppes, P, Winer, M., An axiomatization of utility based on the notion of utility differences, *Management Science*, (1955), 1, 259-270.

[153] Tabucanon, M.T., *Multiple criteria decision making in industry*, Elsevier, (1988).

[154] Ting, H.M., Aggregation of attributes for multiattributed utility assessment, MIT, Operations Research Center, *Technical Repport* 66, (1971).

[155] Trémolières, R., *Partager le pain*, Editions Robert Laffont, Paris, (1975).

[156] Tversky, A., Intransitivity of preferences, *Psychological Review*, (1969), 76, 31-48.

[157] Tversky, A., Kahneman, D., "Cadrage" des décisions et psychologie du choix, *Innovation*, Special Issue 154, (1982), 16-20.

[158] Vanderpooten, D., The interactive approach in MCDA: A technical framework and some basic conceptions, *Mathematical and Computer Modelling*, (1989a), 12, 1213-1220.

[159] Vanderpooten, D., The use of preference information in multiple criteria interactive procedures, in A.G. Lockett and G. Islei (eds.), *Improving Decision Making in Organisations*, Springer-Verlag, (1989b), 390-399.

[160] Vanderpooten, D., The construction of prescriptions in outranking methods, in C.A. Bana e Costa (ed.), *Readings in Multiple Criteria Decision Aid*, Springer-Verlag, Berlin, (1990), 184-215.

[161] Vanderpooten, D., Three basic conceptions underlying multiple criteria interactive procedures, in A. Goicoechea, L. Duckstein, S. Zionts (eds.), *Multiple Criteria Decision Making*, Springer-Verlag, (1992), 441-448.

[162] Vanderpooten, D., Vincke Ph., Description and analysis of some representative interactive multicriteria procedures, *Mathematical and Computer Modelling*, (1989), 12, 1221-1238.

[163] Vansnick, J.C., Strength of preference – Theoretical and practical aspects, in J.P. Brans (ed.), *Operational Research '84*, Elsevier Science Publishers B.V. (North-Holland), Amstardam, (1984), 449-463.

[164] Vansnick, J.C., Measurement theory and decision aid, in C.A. Bana e Costa (ed.), *Readings in Multiple Criteria Decision Aid*, Springer-Verlag, Berlin, (1990), 81-100.

[165] Vincke, Ph., Preference modelling: A survey and an experiment, in J.P. Brans (ed.), *Operational Research '81*, North-Holland, (1982), 341-354.

[166] Vincke, Ph., *Multicriteria decision aid*, John Wiley and Sons, Chichester, (1992).

[167] von Bertalanffy, L., *General systems theory – Foundations, development, applications*, Penguin University Books, (1973).

[168] von Neumann, J., Morgenstern, O., *Theory of games and economic behavior*, 3rd edition, John Wiley and Sons, New York, (1967).

[169] von Winterfeldt, D., Edwards, W., *Decision analysis and behavioral research*, Cambridge University Press, Cambridge, (1986).

[170] Wallenius, H., Wallenius, J., Vartia, P., An approach to solving multiple criteria macroeconomic policy problems and an application, *Management Science*, (1978), 24, 1021-1030.

[171] Wallenius, J., Interactive multiple criteria decision methods: An investigation and an approach, The Helsinki School of Economics, Series A: 14, (1975).

[172] Walliser, B., *Systèmes et modèles – Introduction critique à l'analyse de systèmes*, Editions du Seuil, Paris, (1977).

[173] Wallsten, T.S., Budescu, D.V., Encoding subjective probabilities: A psychological and psychometric review, *Management Science*, (1983), 29, 151-173.

[174] Watson, S.R., The presumptions of prescriptions, Research Paper No. 5/91, (1991), Engineering Department, Cambridge University, Cambridge.

[175] Watzlawick, P., *How real is real? Confusion, disinformation, communication*, Random, (1976).

[176] Watzlawick, P., *The invented reality – How do we know what we believe we know? Contribution to constructivism*, Norton & Company, Inc., New York, (1984).

[177] Weinstein, A.A., Individual preference intransitivity, *Southers Economic Journal*, (1968), 34, 335-343.

[178] Wymore, A.W., *Systems engineering methodology for interdisciplinary teams*, John Wiley and Sons, New York, (1976).

[179] Zionts, S., Decision making with multiple objectives: Some analytic approaches, in J. Ijiry and A. Whinston (eds.), *Quantitative Planning and Control*, Academic Press, New York, (1979), 59-68.

[180] Zionts, S., A multiple criteria method for choosing among discrete alternatives, *European Journal of Operational Research*, (1981), 7, 143-147.

[181] Zoller, M.G., Beguin, H., *Aide à la décision – L'évaluation des projets d'aménagement*, Economica, Paris, (1992).

[182] Zollinger, M., L'analyse multicritère et le risque de crédits aux entreprises, *Revue Française de Gestion*, (1982), 56-66.

[183] Zopounidis, C., A multicriteria decision-making methodology for the evaluation of the risk of failure and an application, *Foundations of Control Engineering*, (1987), 12, 45-64.

INDEX

Nonconvex Optimization and Its Applications

1. D.-Z. Du and J. Sun (eds.): *Advances in Optimization and Approximation.* 1994. ISBN 0-7923-2785-3
2. R. Horst and P.M. Pardalos (eds.): *Handbook of Global Optimization.* 1995 ISBN 0-7923-3120-6
3. R. Horst, P.M. Pardalos and N.V. Thoai: *Introduction to Global Optimization* 1995 ISBN 0-7923-3556-2; Pb 0-7923-3557-0
4. D.-Z. Du and P.M. Pardalos (eds.): *Minimax and Applications.* 1995 ISBN 0-7923-3615-1
5. P.M. Pardalos, Y. Siskos and C. Zopounidis (eds.): *Advances in Multicriteria Analysis.* 1995 ISBN 0-7923-3671-2
6. J.D. Pintér: *Global Optimization in Action.* Continuous and Lipschitz Optimization: Algorithms, Implementations and Applications. 1996 ISBN 0-7923-3757-3
7. C.A. Floudas and P.M. Pardalos (eds.): *State of the Art in Global Optimization.* Computational Methods and Applications. 1996 ISBN 0-7923-3838-3
8. J.L. Higle and S. Sen: *Stochastic Decomposition.* A Statistical Method for Large Scale Stochastic Linear Programming. 1996 ISBN 0-7923-3840-5
9. I.E. Grossmann (ed.): *Global Optimization in Engineering Design.* 1996 ISBN 0-7923-3881-2
10. V.F. Dem'yanov, G.E. Stavroulakis, L.N. Polyakova and P.D. Panagiotopoulos: *Quasidifferentiability and Nonsmooth Modelling in Mechanics, Engineering and Economics.* 1996 ISBN 0-7923-4093-0
11. B. Mirkin: *Mathematical Classification and Clustering.* 1996 ISBN 0-7923-4159-7
12. B. Roy: *Multicriteria Methodology for Decision Aiding.* 1996 ISBN 0-7923-4166-X

KLUWER ACADEMIC PUBLISHERS – DORDRECHT / BOSTON / LONDON

GPSR Compliance

The European Union's (EU) General Product Safety Regulation (GPSR) is a set of rules that requires consumer products to be safe and our obligations to ensure this.

If you have any concerns about our products, you can contact us on ProductSafety@springernature.com

In case Publisher is established outside the EU, the EU authorized representative is:

Springer Nature Customer Service Center GmbH
Europaplatz 3
69115 Heidelberg, Germany

Batch number: 10761599

Printed by Printforce, the Netherlands